DEAD SEXY

How an arms race with parasites
gave us sexuality and mortality

Copyright (c) 2013 by

I0471355

Thomas Whitehead

Dead Sexy is the first of five books in the series

"Viral Schemes Meet Puppet Dreams"

Page 2

Table of Contents

Foreword

Recent years have witnessed some surprising discoveries about the way diseases have affected living things. It seems the struggle between host animals and their parasites has influenced the process of evolution much more than we thought. Scientists have re-examined several of the most puzzling features of life – things that are quite important to us as human beings – and have found that it makes sense to view them as ways of fighting parasites. Our immune systems, our sexual functioning, and our own mortality are among the things we are able to see more clearly in this new light.

The relationship between parasites and immunity is not hard to grasp. But what do parasites have to do with sex and death? Everything, as it turns out.

Sex and death are two of the most dramatic issues in human life. Their emotional impact is powerful. It is strong enough that many of us find it hard to look at them objectively. We recognize the act of sex as the way to have babies. We accept sexual drive and attraction as potent forces shaping our personal identities, our relationships, and our culture. Because we are so close to them we tend not to question them. Not many of us are aware that scientists regard sexual reproduction as a true mystery. Historically they have been unable to fully explain the purpose of sex. What's it all about, really?

Likewise aging, senescence, and dying are things we generally accept without question or examination. But why do we have to die? Is death truly inevitable, or are there ways to dodge or postpone it? Why do we have to "get old?" Do our energy, health, and mental clarity have to dwindle away to the vanishing point as we age? Do our bodies have to fall apart?

This is a book for people who want answers to these questions. It tells a remarkable story. It is the true tale of a never-ending war – an arms race between host organisms (like us) and the parasites that exploit them. This life-or-death struggle has been going on for an inconceivably long time, long enough to allow host organisms to evolve staggeringly intricate measures for self-protection. The surprising truth is that both sex and death are inventions, weapons developed in the course of the arms race with our parasites. My intent is to explain why these weapons were invented, and how they protect the animals that created them. Better understanding will lead to some practical suggestions for improvement in length and quality of life.

Chapter 1 casts a penetrating light upon parasites and their ways. It clarifies what they are and how they operate. It explains how animals (including us) were forced into an ongoing battle with them. And it highlights an issue of special concern: parasites' manipulative tendencies.

Chapter 2 examines the immune system. The biological functions we place under the umbrella of immunity are an intricate collection of defensive weapons. Their purpose is to stymie parasites. This chapter looks at why these defenses evolved, and provides a rough outline of how they work.

Chapter 3 explains why sexual reproduction should also be regarded as a defensive weapon, and places it under the larger umbrella of immunity. It tells where sex came from, how it fits into the picture of life's evolution, and how it works to combat parasites. It also details some surprising things about the real reasons we are sexually attracted to some people, and not others.

Chapter 4 places death under the same microscope. Programmed death is the expected end of life that is normal for humans and most other complex animals. The available data suggest that senescence and death from old age are initiated by the animal itself. This brand of mortality was invented in the course of evolution for the best of reasons, and is carried out under our own genetic control.

Chapter 5 presents some practical applications of the information in this book. It details a number of scientifically validated methods that readers can start using now to extend their lifespans, and improve the quality of their lives.

In putting this book together I drew scientific information from multiple sources, and have provided citations for many of them. Of course, people have different viewpoints and don't always agree. The ideas expressed here are accepted by many mainstream scientists, but certainly not by all of them. Researchers can interpret the very same data in widely different ways. In the end, I would claim only that the information in the book is based upon the views of a large group of responsible researchers and theoreticians. I present them for your consideration.

Dead Sexy is the first book in a series of five. Each of the books can stand on its own. But they are designed to fit together like the pieces of a jigsaw puzzle to create a larger picture. The series is entitled "Viral Schemes Meet Puppet Dreams: Behavioral parasites diminish our awareness, robbing us of knowledge, purpose, and direction." You'll find both the Series Foreword and the Acknowledgments at the end of this book.

Thomas Whitehead

Chapter One
Few boats, lots of passengers

In life there are producers and there are exploiters. The producers are those who actually create things. The exploiters don't create much themselves. Instead they grab things that the producers have made, and use them to their own advantage. The exploiters – we call them parasites – live off the bounty of their unwilling hosts.

A group of reputable scientists have recently concluded that sex and death are inventions. They are products of an ongoing struggle between parasites and the animals they exploit. The present chapter clarifies the nature of that struggle, and so sets the stage for an understanding of why and how these inventions came to be.

It goes without saying that there are some genuine advantages to being a parasite. It's much easier to ride in a boat than it is to row it. Parasites creep aboard their host organisms, setting themselves up for a free trip down the river of life. So parasitism is popular. The majority of all known species are parasitic, from viruses to much more complex life forms.[1] Author Carl Zimmer, in his fascinating book *Parasite Rex* writes

Every living thing has at least one parasite that lives inside it or on it. Many, like Leopard frogs and humans, have many more. There's a parrot in Mexico with 30 different species of mites on its feathers alone. And the parasites themselves have parasites, and some of those parasites have parasites of their own.... According to one estimate, parasites may outnumber free living species four to one. In other words, the study of life is, for the most part, parasitology.[2]

From the earliest glimmerings of life upon this earth, hard-working, boat-rowing organisms have had no choice but to spend precious time and energy defending themselves from these pushy stowaways.[3]

Kinds of parasites

We put parasitic organisms into two classes – simpler disease entities such as viruses and bacteria, and more complex creatures with much more sophisticated capabilities. The distinction between the two is arbitrary, and peculiar to American and European culture. There is no actual dividing line between disease entities and parasites.[4] These pathogens are everywhere. No animal is immune. Ecosystems typically include a whole network of host/parasite relationships. The details of their complex interactions are only now beginning to be unraveled.[5]

Some parasites are quite simple, others complex. Some are animals, some are plants. Some work through chemicals, others use behavior to squeeze resources from their host.

Here's an example. The common cuckoo (scientific name *Cuculus canorus*) is a "brood parasite." It lays a single egg in the nest of a smaller bird. The cuckoo chick usually hatches first. Soon afterward the cuckoo changeling "ejects the hosts' eggs by balancing them on its back, one by one, and heaving them over the nest rim. So the cuckoo gets the nest to itself and the hosts then slave away defeated for five weeks, even as it grows to 10 times their own body weight."[6] [7]

With false visual, chemical and behavioral cues, the cuckoo chick commandeers the caretaking efforts of its stolen parents. It does this by taking advantage of the way evolution has programmed the host parents. Some cuckoo species make their eggs look almost exactly like the eggs of their hosts. They "have evolved such good host egg mimicry that the only way for humans to distinguish the cuckoo egg is by its thicker shell."[8] Common cuckoos have developed ways to extract the maximum food and caretaking from their hijacked foster parents by "pressing their buttons." For example, the single cuckoo chick tricks the parents into providing multiple helpings of food by making rapid calls that sound like the chirping of many of their own young.[9] The parents behave as if they are dutifully providing for a whole nest full of their own chicks. The cuckoo hijacks the complete reproductive effort of the host parents, subverting it to create the next cuckoo generation instead.

Not all cuckoos are parasites. Of the 141 species in the cuckoo family, 82 raise their own young, while 59 are brood parasites.[10]

An example of a parasitic plant is the Dwarf mistletoe (genus *Arceuthobium*). Species of this plant grow attached to several types of conifer in Western North America and around the world. Tendrils from the mistletoe plant penetrate into the tissue of the conifer. Then, like a botanical mosquito, it siphons off nutrients. Dwarf mistletoes acquire all of their water and nutrients from their hosts. They can't survive on their own. They are called "obligate" parasites because they can't stay alive without their host.[11]

Biological viruses are also parasites. Viruses are labeled obligate intracellular parasites.[12] "Intracellular" because they do what they do only inside living cells. Viruses are able to replicate only by exploiting the replication machinery that's inside the living cells of organisms. Viruses are everywhere. They are diverse in their characteristics, infecting all types of cell. Their hosts include both complex animals (like us) and single-celled organisms. Bacteria, blue-green algae, fungi, plants, insects, and vertebrates (animals with a backbone) are all viral hosts.[13]

A new world of wonder

It seems Darwin didn't know enough about parasites when he conceived the theory of evolution. In his day they hadn't been studied extensively. Darwin himself appears to have been rather put off by their heartless, extractory ways. He apparently considered them only in passing. Darwin gave them attention generally when he was trying to reconcile his concept of a benevolent God with the base behavior of certain parasites, dastardly animals that (disturbingly) were created by that same God.[14] His inner turmoil is reflected in an early draft of his theory.[15] There he seems to resolve his distaste for the image of God-the-parasite-creator through his insight that natural selection was the direct creator of these villains. Darwin grasped that it was natural selection, a purely mechanical process, that produced these base creatures – not God. So natural selection lets God off the hook, so to speak. With apparent relief, Darwin wrote

> It is derogatory that the Creator of countless systems of worlds should have created each of the myriads of creeping parasites and worms which have swarmed each day of life on land and water on one globe. We cease being astonished, however much we may deplore, that a group of animals should have been directly created to lay their eggs in bowels and flesh of other [as parasitic worms and wasps do], –that some organisms should delight in cruelty, –that animals should be led away by false instincts....[16] [17]

We're living in a different time now. We no longer feel compelled, as did Darwin and his contemporaries, to debate the morality of parasites. Greater objectivity has opened the gates to a whole new world of scientific wonder. A mother lode of mind-boggling information has been unearthed, letting us better grasp the true influence of parasitic lifeforms. We are beginning to appreciate their considerable impact upon evolution itself. The news isn't all bad. We are realizing that many of the most awe-inspiring creations of living things originated as defenses against parasites.

The drama of predator and prey

"Survival of the fittest" is a phrase that has been used as a shorthand description of evolution. Consider the drama of the African lion and its prey the wildebeest, locked in a conflict spanning eons. In times past we saw the Darwinian struggle for survival in the interaction of these two species. It was a contest that drove the evolution of both prey and predator. We understood that the lions went after the wildebeest that were less fit – and so easier to catch. These were animals sicker, or weaker, or slower than the others. Over time, we believed, the less fit prey were weeded out. That left the strong to contribute their genes to future generations. That's natural selection in action. For their part the unfit lions – the clumsy and inept ones who missed too many meals – were also weeded out. That left their more capable fellows to reproduce. Survival of the fittest indeed. That seemed a credible enough accounting of the forces driving evolution.

Well, there was a lot of truth in this accounting. But not enough truth, as it turns out. We now know that the picture was far from complete because it didn't include the parasites. Each of the players in the lion/wildebeest drama does have its evolutionary impact. But it seems that the player most influencing evolution may have been the one that has historically drawn the least attention. That player has been acting behind the scenes. It is the parasite inhabiting both predator and prey.

Heteroxenic parasites

Ectoparasites are animals like ticks and fleas that attack their hosts from the outside ("ecto-" from the Greek "ektos" meaning outside). Endoparasites like liver flukes and intestinal worms attack from the inside ("endo-" from the Greek "endos" meaning within). In the wild, all animals with backbones are vulnerable to ectoparasites and endoparasites. But the parasites are much harder to see than their host animals. So they have been neglected by ecologists, even though their sheer mass makes up a big part of the ecosystem.[18]

One class of endoparasite lives within a single host during its entire lifecycle. This type is called "monoxenic" ("mono-" from the Greek "monos" meaning single; "xenic" from Greek "xenos" meaning foreigner). But the lifecycle of another type of endoparasite – an important one – is more complex. The "heteroxenic" parasite grows inside more than one other animal ("hetero-" from the Greek "heteros" meaning other). It moves through two or more hosts in succession. A heteroxenic parasite generally first lives in a host where it multiplies by simple splitting (this is called the intermediate host). Then it moves to a second animal (called the definitive host) where it reproduces sexually.

The trip from the intermediate host to the definitive host takes place when the definitive host eats the intermediate host. For example, a rat (intermediate host) commonly contains heteroxenic parasites that travel to the cat (definitive host) that makes a meal of the rat. This is called trophic transmission (after the Greek "trophos" meaning pertaining to nourishment).

No waiting

We might imagine that the little heteroxenic parasite patiently bides its time until its intermediate host is eaten. Sort of like waiting for a bus. But that really isn't the case. It seems many heteroxenic parasites exert substantial control over their host animals. Their influence isn't subtle; it's not benign either. Heteroxenic parasites are often able to manipulate their hosts' behavior. They are able to control the activity of the prey animal well enough to offer it up as food for the predator animals. That's a thought spooky enough to be science fiction. But it's real. How do they do it?

The extended phenotype

Animals have a genotype and a phenotype. The genotype is the collection of information in the organism's genes. A rough analogy would be the architect's blueprint for a house. The phenotype is the way those genes are expressed in the form of the body and behavior of the animal. So the phenotype would be the house as it turns out when the contractors and decorators have finished their work.

As famously explained by writer Richard Dawkins[19] animals have not only a phenotype, but an "extended phenotype" as well. This is the set of characteristics expressed *beyond* the body of the animal. For example, a beaver's dam and a rabbit's burrow are part of their extended phenotypes. Part of the extended phenotype of the heteroxenic parasite is the behavior of its intermediate host. In other words the genes of the parasite partly control the behavior of the host.

It may seem strange that one animal could express its genes through the behavior of a second animal. But that's how it is with parasites. Although the prey animal's behavior certainly belongs to itself, some part of that behavior also belongs to the parasite inside it. When the wildebeest moves, it is certainly the wildebeest moving. But it's also the parasite moving. An unsettling idea, and one of the reasons why movies like the 1956 *Invasion of the Body Snatchers*[20] and the 1979 *Alien*[21] (in which parasitic aliens invade humans and use them as host animals) both captivated and creeped out so much of their audiences.

Just how does a heteroxenic parasite take control of its host? Researchers Fenton and Rands discuss their influence.

The most striking examples involve trophically transmitted parasites which manipulate their intermediate hosts to make them more vulnerable to predation by the parasites' definitive hosts. These manipulations can occur through increasing conspicuousness or by altering the behavior of the intermediate hosts to make them easier to be captured. Such manipulations of host behavior are fascinating examples of the extended phenotype of the parasite.[22]

One heteroxenic parasite family is called *Sarcocystidae* ("sarco-" from the Greek "sarx" meaning flesh; "cyst-" from the Greek "kystis" meaning bladder or sac). Researchers Seilacher et al write

The Sarcocystidae, in particular, are characterized by forming cysts and by changing between two warm-blooded hosts (mammals, birds) during their life cycles... In every case, one of the hosts serves for the asexual multiplication (intermediate host, by definition), the other for the sexual multiplication (definitive host). Sarcocystidae do less visible harm to their intermediate (herbivorous) hosts than to the definitive (carnivorous) ones. Nevertheless, they are able to alter the behavior, particularly of the intermediate host, in the interest of the parasites' transmission (bait strategy). This bias is reflected in the fact that infected individuals are more common in the kill than the rest of the prey population.[23]

Seilacher et al describe the interaction of wildebeest and their natural predators, African lions. Wildebeest serve as intermediate hosts for a type of *Sarcocystidae*, with the lions being the definitive host.

... [In] spite of being aware of the stalking lions, they do not run away in stampede because they "know" that the selection has already been made: During their initial inspection, the lions have

decided for an individual that showed some kind of risky behavior, such as grazing at a distance from the herd. Thus the "bait" laid out by the parasite has already been effective.[24]

Managing the food supply

But really, how much impact could parasites actually have on the food chain? More than anyone ever dreamed possible! Recent studies suggest that in many ecosystems the *vast majority* of the food entering the mouths of predators was put there by parasites within the prey. Researchers Lafferty et al studied food webs in a salt marsh ecosystem. They remark that a common snail parasite (scientific name *Euhaplorchis californiensis*) uses a fish as an intermediate host. "The parasite larval *E. californiensis* encyst on the fish's brain and manipulate behavior, rendering infected fish 10–30 times more likely to be eaten by one of the 15 bird species ... that serve as a final host for the adult worm."[25] In this case the parasites' control of the fish seems pretty good – good enough to multiply its chances of being eaten up to 30 times above that of an uninfected fish. It's almost as if the parasites were making the fish jump directly into the mouths of their predators.

So parasitic manipulation of hosts can be quite effective. It is also flexible. The door can swing both ways. Before parasitic cysts have matured within the intermediate host – before the cysts are ready to be eaten – they often exert a protective influence. That is, they change their hosts' behavior to make them *less* likely to be victims. But once the cysts are mature and ready for transmission to the definitive host, they make them *more* likely to be eaten. They cause changes in behavior that transform the hapless animal into predator bait. The host ends up behaving in ways that run directly against its own interests.

Focus on heteroxenic parasites

Back to the saga of the wildebeest and the lion. While they are still inside the wildebeest, the heteroxenic parasites reproduce asexually by splitting. They form cysts in the wildebeest's flesh – in both muscle and brain. Each of the cysts houses a large number of individual parasites named zoites (from a Greek word meaning living thing). While they are within the cysts they are called cystozoites. Each individual zoite has the potential to infect a lion that happens to eat the tissue. But – and this is a critical point – the cystozoites aren't identical.

As part of their life strategy, cystozoites reproduce within the prey animal by splitting in a way that creates a lot of mutations. Each cyst ends up containing zoites with a lot of variations between them. These mutations are no accident. They are an essential part of the parasite's strategy. As we will see, host animals are always trying to shake off their parasites through various defensive maneuvers. The mutations create enough diversity within each new generation of parasite to allow it to keep up with the evolving resistance of its host. Seilacher et al say

> *During the long residence in the herbivorous hosts' tissues, mutations have a chance to accumulate among the thousands of cystozoites enclosed in a single cyst. These mutations become subject to sexual recombination in the definitive (carnivorous) host. By favoring small-scale evolutionary changes, heteroxenic cycles thus provide the parasites with the genetic diversity they need in the interaction with the hosts' defense systems.*[26]

When the lion eats tissue containing a cyst, the individual zoites pop out. They are now free to move around inside the lion. In this form they are called trophozoites. Inside the lion they find mates and reproduce sexually. They package a new generation of parasite eggs into small oval containers. These packages leave the lion's body with its feces. Deposited into the grass, some of the containers will be consumed by grazing wildebeest to start the cycle all over again.

But the feces-ensconced eggs must clear one final hurdle. Over countless generations of infection, wildebeest (displaying wisdom common to most other grazing prey animals) have come to instinctively avoid recognizable piles of feces. They stay away from them even if the feces are very old. So the wildebeest's simple tactic of avoiding lion poop could potentially stop the parasite in its tracks. Unfortunately for both wildebeest and lion the parasite has one final trick up its heteroxenic sleeve.

Onto the scene march "transport hosts." These are insects whose only function in the parasite's reproductive cycle is to spread the egg-riddled poop around. Insects that make a living from the lions' feces break it up into much smaller pieces and spread it over the grass. These tiny pieces can't be seen or avoided by the wildebeest.[27]

Opportunistic Evolution

The parasites clearly benefit when they cause wildebeest to sacrifice themselves to the lions. But exactly how are tiny, brainless parasitic organisms able to control a wildebeest's behavior precisely enough to pull this off?

The answer lies in the opportunistic nature of evolution. The parasites incorporate into their reproductive strategy whatever by chance happens to help them reproduce – and reproduction happens after the host is taken by its predator. There are many different things that affect the likelihood of a prey animal's being eaten. Some of these are quite subtle. It could be something as simple as weakening the animal, or slowing its reaction time, or making it less coordinated – any of these things reduce the likelihood that it will outrun its predator. Or it could be accomplished by reducing the level of fear that the prey animal experiences, making it less cautious and more likely to take risks. Or it could be changing the animal's behavior so as to attract more attention. The way evolution works, it could be *anything at all* that gets more wildebeest eaten. The high level of mutation in the parasites increases the odds that they will end up pressing a useful button by chance. Over time the winning strategies are systematically collected into the parasite's host control toolbox. Finally what began as a random event is shaped into a well-defined strategy.[28]

For example, one heteroxenic parasite uses as its intermediate host a tiny shrimp-like creature called an amphipod (*Gammarus roeseli*). By releasing chemicals within the amphipod's body the parasite is able to change the way the little animal responds to light. Uninfected amphipods naturally move away from the light toward the depths of the ponds where they live. This protects them from predators on the surface of the pond. Infected amphipods, by contrast, move toward the light – to the surface of the pond. This simple change makes them easy targets for ducks, the parasite's definitive host.[29]

In the line of fire

Another example of the control exerted by a heteroxenic parasite is more immediately relevant to human beings. It is the disease toxoplasmosis, caused by a heteroxenic parasite with the scientific name *Toxoplasma gondii*. The definitive hosts are members of the cat family, including common house cats. The intermediate hosts are small rodents such as mice and rats. Unfortunately, humans commonly get infected with this parasite as well.

Here's some bad news: It is estimated that between one third and one half of the world's human population is infected with *T. gondii*.[30] [31] That is, the cysts of these parasites are in our flesh and brains too. Although humans are not the actual intermediate host (people are not commonly eaten by cats), they do appear to be largely responsible for the success of *T. gondii*. Carl Zimmer notes that humans helped *T. gondii* by spreading cats (intentionally, as pets) and rats (unintentionally, as pests) around the world.[32] It is estimated that in the US about one person out of every five in the population is infected. It's not hard to see a potential problem here.

Like other heteroxenic parasites, *T. gondii* can change the behavior of its intermediate hosts. When the host is a mouse or rat, these changes make it more likely that the animal will be caught and eaten by a cat. Infected mice show increased levels of activity and aggression. And the parasite changes their behavior in other ways.

Researchers Ferguson and Hutchison did detailed microscopic examination of infected rat brains to see just what was going on under the hood. Their study revealed that "all the cysts were present within intact host cells irrespective of their size or the period PI [post-infection]. The majority of host cells could be positively identified as neurons..." In other words, each cyst was found nestled directly within a living individual brain neuron. The authors believed that the cysts are hidden inside neurons to protect them from recognition and attack by the host immune system.[33]

The odor of cat urine normally strikes fear into the hearts of rodents. But infected rats seem to lose their natural fear of cat odors. In fact, they seem to find the odor of cat urine interesting and stimulating – maybe even sexually stimulating. Infected rodents have increased levels of the neurotransmitter dopamine in their brains, leading to "novelty seeking and neurotic behavior." They end up hanging around in just the places they should at all costs avoid – places they can smell cats. These things interfere with the rodents' ability to defend themselves, increasing the likelihood they will be eaten.[34]

Toxoplasmosis clearly does change the behavior of rats, and

that's no accident. It's part of the parasite's life strategy. And people get infected too.

According to Australian researcher Nicky Boulter, "infection generally happens by accidentally eating oocysts (the parasite's equivalent of eggs) excreted in the faeces of an infected cat ... or, most commonly, by ingesting cysts containing the parasites in raw or undercooked meat."[35]

The parasite is clearly a danger to pregnant women as it can cause disability or abortion of an unborn child. But can it change humans' behavior as it does with rats? Unfortunately multiple studies have confirmed that it can and does change human behavior.

According to Boulter, "the effect of infection is different between men and women." Women infected by *T. gondii* seem to become more outgoing, friendly, conscientious, kind – and more attractive to men. But "infected men have lower IQs, achieve a lower level of education and have shorter attention spans. They are also more likely to break rules and take risks, be more independent, more anti-social, suspicious, jealous and morose, and are deemed less attractive to women." Boulter sums up the changes this way: "In short, it can make men behave like alley cats and women behave like sex kittens."[36]

It might seem that men are getting the worst of this. But both infected men and women are more prone to feelings of guilt.[37] A 2009 study suggested that infection may triple the likelihood of driving accidents.[38] Infection appears to decrease novelty seeking in both men and women. And preliminary findings hint that toxoplasmosis may be associated with the development of serious mental disorders including schizophrenia.[39] [40]

It has been shown that the *T. gondii* genome codes for two chemicals that could directly affect the body's ability to make dopamine and/or serotonin – both neurotransmitters known to heavily influence human behavior.[41] These chemicals have been custom-tailored by evolution to affect the brains of its true intermediate host, the rodent. The chemicals were evolved opportunistically. Some of the cystozoite mutants just happened to produce chemicals with some behavioral impact on the rats, and the next generation produced variants even more effective, and so on.

As a parasite, *T. gondii* is shooting at the rats, not at us. Unfortunately we are in the direct line of fire. The behavioral changes they induce in humans may not be the same ones the parasite finds so useful in rodents. But they're real anyway.

Monoxenic parasites too

Clearly, heteroxenic parasites can and do alter the behavior of their hosts. Can monoxenic parasites pull off the same kind of trick? It seems that they can.

Parasites of the water flea *Daphnia magna* have been studied in some detail. Unlike their heteroxenic cousins, most monoxenic parasites perish when their hosts are eaten. Researchers Fels et al determined that monoxenic parasites of the water flea influence it to stay deeper in the water than its uninfected fellows.[42] Presumably staying at a greater depth protects *Daphnia* and its parasites from predators at the surface.

Hairworms (*Spinochordodes tellinii*) are long, stringy organisms that can be found in streams, puddles, watering troughs, etc. They're generally 20 to 40 inches in length, but can become much longer. As adults they are free-swimming. But the larvae are parasitic and live in insects. Crickets and grasshoppers normally don't like to get into water. But when infected with hairworm larvae they display some decidedly strange behavior. The unfortunates dive headfirst into a body of water. The parasite then bursts out of the hapless insect just like a creature from the movie *Alien*, killing the host. Freed from the host body, the hairworm settles into its new home in the water. It finds a mate, reproduces, and so starts the cycle all over.[43] [44] The hairworm is a hardy little fellow. Remarkably, if the host grasshopper should happen to be eaten by a bird, the worm is often able to save itself by wiggling out of both host and bird.

Because of the opportunistic way parasites gain their influence, it is likely that every parasite (monoxenic and heteroxenic alike) will gradually gain some measure of influence over its host's behavior. Why? Because any "host button" the parasite can press to promote its own survival, it will eventually press by chance over the generations. Those parasites that press the right buttons will survive more often, and pass their skills on to their progeny. The form their influence takes will vary. But there's little doubt it will be something that enhances the survival and reproduction of the parasite.

In summary

Information about parasites' ability to control hosts has dramatically changed our view of predator/prey interaction. We now know that in many ecosystems, the apparent struggle between prey and predator animals is scripted. Despite appearances, the larger animals are merely actors in a drama they can't control. The parasites wrote the script, and it's the parasites directing the action. It's the parasites who decide which individual prey animal gets eaten by predators. The parasites determine which organisms proliferate, and which do not. As unlikely as it may seem, these tiny brainless critters are able to bring bigger, smarter, stronger, more complex animals under their control.

Why do they do this? Parasites are simple animals with limited abilities. They use their host animals' much more extensive abilities to arrange the parasites' reproduction. In a way it's like flowering plants and their helpers. Plants use insects and birds to carry their pollen around – something they can't do themselves. But the plants pay their helpers with nectar in what seems a fair trade. Parasites generally don't compensate their hosts for what they take.

How do parasites gain the ability to influence their hosts? The answer lies in their opportunism. Their method of reproduction insures a high level of random mutation. Some of the mutants happen by chance to influence host behavior in a way that helps the parasite. These mutants are more likely to survive and pass genes on to the next generation. So over time the parasite is able to exert greater and greater control over its host.

The previously neglected struggle between host animal and parasite appears to have deeply influenced the evolution of species. Because the influence of parasites is profound, their influence on evolution is also profound. Some of the most mystifying things about animals can be understood as extreme measures for coping with parasites. We will take a look at some of those measures in the chapters that follow.

Chapter Two
The immune system

Do host animals go along helplessly with parasitic manipulations, or do they somehow fight back? Of course they fight back. They *must* fight back. If they didn't, they wouldn't be around for long. Parasites' free ride is not without cost to the host organism. Supporting freeloaders diverts resources from the host, resources it needs to stay healthy. Such a cost places infected individuals at a disadvantage. It favors the survival of organisms better equipped to deal with it. Over the ages this steady selection pressure has led to the evolution of some brilliant defensive mechanisms. Among the most stellar of these counter-maneuvers is biological immunity.

As animals began to develop immune defenses, the parasites were obligated to find ways to defeat or sneak around those defenses. This led to a so-called "arms race." Each side had to fight the other just to stay alive. It is the arms race that led to the most elaborate of the measures taken by host animals – including sex and programmed death.

In this chapter we'll consider what immunity is, how it came to be, and why the arms race was inevitable.

Boundary keeping

Immunity protects the self from harmful outside influences. Self-protection makes no sense unless there is a self to protect. The cell membrane defines the self of the cell. It is a boundary that separates the internal workings of the cell from what is going on outside. By enabling the cell to start creating order inside itself the cell membrane ramped up the pace of evolution.

In the natural world disorder increases over time. That's called entropy. But life defies entropy, and hangs onto order. And over time, miraculously, it even increases order.

The law of entropy says the universe is running down. All closed physical systems are moving toward disorder and chaos. So if you put a box around any area of space, and don't let anything in or out, things inside the box will get messier and messier with time.

It is true that in a closed physical system entropy always wins out. But as deep thinker Ludwig von Bertalanffy long ago pointed out, living systems are not closed. And because they aren't closed they're able to cheat. Life persists by dodging entropy's inexorable grind. How? Living things channel disorder away from themselves; they push it outside their boundaries. "Thus," says von Bertalanffy, "living systems, maintaining themselves in a steady-state, can avoid the increase of entropy, and may even develop towards states of increased order and organization."[45]

Life is all about repeating patterns. Since it first began life has increased tremendously in complexity, but it always involves forms reproducing themselves. Reproducing isn't easy without a controlled environment. Only the simplest organisms trust nature to consistently provide conditions favorable to them. More complex organisms define a space, then develop ways of controlling conditions inside that space. That's how they beat entropy. Such maintenance processes are characteristic of living things. They have been in evidence from the very beginning of life's history.

Managing entropy with boundaries has the quality of intelligently directed action. The intelligence flows not from a brain, but from evolutionary pressures. It is wisdom accumulated over the eons through natural selection.

Managing boundaries is so central a feature of living things that it appears again and again over the entire spectrum of life, from the simplest one-celled organism through multicellular organization, through animal behavior – all the way to culture, law, and principles of nationhood.

Here's an example: At the level of behavior we witness animal territoriality. Many species mark off territories for themselves, detect intrusions, and aggressively expel intruders. They exert control over events inside the boundary they have defined for themselves.

Another example: Each of us has boundaries around our personal space, systems that define ownership. Our boundaries let us know when we are being violated, and prompt us to action to defend ourselves. Each of us has boundaries around our sexual selves, defining for us who we may touch (and be touched by) sexually, and under what circumstances. More universally, our boundaries define who we are, alerting us to impulses inconsistent with our self-image. This boundary is our notion of "who we are." It prompts measures to control or eliminate unacceptable impulses. Because boundary systems exist in all aspects of human life we are immersed in them, to the point that they are almost invisible.

The roots of awareness

Life is about surviving and passing the formula for survival on to the next generation. Over the course of evolution life has developed what much resembles an "urge" to do the things that promote life. Even a single cell somehow remains "aware" of its chemical state, and takes action to maintain it within certain limits.[46] Author and researcher Antonio Damasio eloquently voices precisely this point in his engrossing book *The Feeling of What Happens*. He writes,

The unwitting and unconscious urge to stay alive betrays itself inside a simple cell in a complicated operation that requires "sensing" the state of the chemical profile inside the boundary, and that requires unwitting, "unconscious knowledge" of what to do, chemically speaking, when the sensing reveals too little or too much of some ingredient at some place or time within the cell. To put it in other words: it requires something not unlike perception in order to sense imbalance; it requires something not unlike implicit memory, in the form of dispositions for action, in order to hold its technical know-how; and it requires something not unlike a skill to perform a preemptive or corrective action. If all this sounds to you like the description of important functions of our brain, you are correct... [S]ensing environmental conditions, holding know-how in dispositions, and acting on the basis of those dispositions were already present in single-cell creatures before they were part of any multicellular organisms, let alone multicellular organisms with brains.[47]

When we talk about simple lifeforms like bacteria, we know they have no mind. Yet we find it difficult to talk about them without using psychological terms. We speak of the "strategy" of viruses, though we are well aware that viruses don't have brains or wills, and so could not possibly have an actual strategy. We say a cell "recognizes" a pathogen and "defends" itself against it. But we know the cell has no visual sense organs, and so can't recognize anything. It can't act, plan or will things as you and I can. So it can't defend the way we could.

It's hard to avoid using psychological words because there's such an obvious connection between what we see the cells doing and what we ourselves do in our everyday lives. When I place quote marks around psychological terms I, like Damasio, am making a statement about the continuity between elementary life and mental processes. It does seem there is a natural progression from the "intelligence" (quotes required) that natural selection brings to cells, to the intelligence of our own brains (quotes no longer necessary).

Housing in a hurricane

It would be hard to build a house in the midst of a hurricane. And it must have been hard for the very earliest elements of life to hold themselves together while floating around (presumably) in the unprotected environment of the primordial sea. There's some evidence that when life first began to stir every nest of living chemicals could share genetic information with every other nest. "Indeed, the absence of barriers to DNA exchange (general genetic mixing) may have characterized the first main stage of cellular evolution.[48] It was a sort of worldwide "living soup."

But cellular forms of life – the forms familiar to us today – really began to take off when bundles of self-reproducing chemicals somehow developed the capacity to surround themselves with a membrane. They created order by walling themselves off from the chaos. That's the way they stood entropy on its head. The enclosing membrane created a tent in the wilderness. Now there was a controlled space where conditions could be regulated, and made friendly to life.

Boundary keeping is a primary means of accomplishing this entropy-undoing sleight of hand. Once the boundary is established, things that don't fit in with life functions can be moved to the outside. The roof and walls surrounding a home keep the rain and snow out. Dirt and insects (and strangers) that happen to find their way inside are detected and removed. Order is created inside the home by throwing all the disorder outside.

Our air conditioner detects excess heat, and then transfers it from the inside of the building to the outside. Heat is not destroyed in the process, of course. In fact the compressor creates even more heat. But this doesn't matter much to us, because all the heat ends up outside our boundary. We're inside where it's cool. Entropy's happening, for sure. But it isn't happening to us, at least for the moment. We've defeated it with boundary keeping.

Life defines a space. Then it detects and gets rid of everything that doesn't belong in that space. Of course, life's judgment of what "belongs" is rooted in requirements for its own survival. It could hardly be otherwise. Various aspects of a cell's internal environment are tightly controlled by evolutionary pressures. For example levels of life-critical chemicals like sodium, potassium, and calcium are kept constant inside the cell. This enables bioprocesses that can only happen within a narrow range of concentration. Order is created inside the cell.

Immunity as boundary keeping

Because they are smaller and simpler, parasites have a built-in edge over their hosts. It is widely appreciated that they can evolve very fast – much faster than their hosts.[49] Left unchecked, parasites would quickly overwhelm and destroy a species. So host animals have needed protection from them. The need for protection created selection pressure. And it's that pressure that, over the ages, has led to the evolution of the immune system. The immune system is a set of boundary keeping mechanisms. It's an elaborate collection of amazingly sophisticated functions for detecting and controlling parasites.[50]

Now take a moment to look at things from the parasite's point of view. Parasites can't afford to let their hosts' immunity defeat them. If they want to live, they have no choice but to develop ways to counteract host immunity. They try to hide from it, overpower it, or inactivate it. So parasites and their hosts stay locked in an eternal struggle, each unable to live without finding ways to stymie the other. It's an arms race. As we are coming to understand, the story of life on this planet is in large part the story of a war – the war between host organisms and the parasites that unceasingly seek to exploit them.[51]

The body has developed several layers of protection from disease organisms. Each layer is a form of boundary keeping. A first boundary is the skin, which blocks the path of micro-organisms that would otherwise be invaders. Though the skin is pretty effective, some pathogens do get past it. When they do, they must overcome other barriers. If they get into the lungs automatic defensive behaviors come into play. Sneezing and coughing knock pathogens and irritants out of the respiratory system. If they get into the eyes, tears flush them out. Further, mucus secreted by the respiratory and gastrointestinal tract bogs invaders down and immobilizes them.

When pathogens get past these, they must still fight past chemical barriers. In the stomach, gastric acid and other special chemicals defend against those entering with food. Cells in the skin, eyes, and respiratory tract secrete special chemicals called "defensins" that punch holes in attacking microbes. After girls begin menstruating their vaginal secretions become acidic, forming an inhospitable environment for most bacteria. In men, semen kills bacteria with both defensins and zinc. And in both men and women friendly bacteria compete with pathogens for food and space, and crowd them out.

Inevitably, some pathogens manage to jump all these hurdles. Meeting these persistent invaders is a final additional complex of protective mechanisms. By tradition, these are collectively called "the immune system."

The job of the immune system is to identify and neutralize potentially harmful invaders. It does its job selectively, sensitively – one might even say "intelligently." Its task is to continually edit pathological patterns out of the animal's normal life processes. Every organism that has to deal with parasites (and that's every organism) has been driven to evolve an immune system. Even bacteria have immune responses. Vertebrate organisms (like us) have developed their immune systems to sublime levels.

To be clear, this high level of development is a direct result of strong, sustained selection pressure. Potential invaders roam about in great numbers. Individual organisms that don't have effective ways to deal with them are very much at risk. More to the point, these less competent individuals die more often. The ones that survive are those better at resisting. These individuals live to pass their genes to the next generation. In this way natural selection strongly spurs every species towards refinement of its capacity for effective immune response.

When it comes to the parasite the shoe is on the other foot. If it is to exploit the host's resources the disease entity is obliged to battle its host's immune system. Parasites that can't deal with the immune system are toast. So a parasite's ability to thwart host immunity is crucial to its fitness.

Host must defeat parasite. Parasite must defeat host. These are the earmarks of an arms race.

Innate immunity

Immune responses come in several different flavors. Even simple organisms have some basic immune functions. They have to. Multicelled organisms have much more sophisticated capabilities. Both vertebrate and invertebrate animals have innate (meaning from birth) immunity, a basic system of defense against potential pathogens. Vertebrates have in addition a more elaborately developed form called adaptive immunity.

The way the *innate* immune response works is like keeping the streets safe by making regular patrols, and indiscriminately rounding up anyone who doesn't seem to belong there. The *adaptive* immune response (which we will examine a little later) is like identifying a specific criminal, then sending out informed officers to locate and arrest him.

Innate immunity has two parts:

1. Humoral Immunity. We call the first part of innate immunity "humoral." That's not because it's funny. It's because its agents are always floating around in the organism's body fluids – the archaic term for which is "humors." In these fluids we find molecules that can slow the growth of pathogens. They can also "tie them up" (clump them together) rendering them harmless and helpless, so they can be flushed out of the body.

2. Cellular Immunity. The second part of innate immunity we call "cellular." That's because its agents are specialized body cells. There are several types of these special cells. Two of these are the macrophages and the natural killer cells. (a) Macrophages are amoeba-like cells that roam the body, "eating" and digesting any foreign bodies they encounter. They aren't picky eaters. Their motive is defense, not hunger. They will attack and devour anything that doesn't seem to belong there. (b) Natural killer cells also roam the hallways of the body. They're called killers because they can act against the body's own cells. They recognize and destroy cells that don't seem to be working as they should – for example those that are infected with a virus or have become cancerous.

These forms of innate immunity – humoral and cellular – are impressive and effective. But they are nonspecific. They are directed against invaders in general.[52] Although they make a brilliant defense, they aren't "smart" enough to target specific types of invaders.

Adaptive immunity

Some pathogens get past innate immunity too. But jawed vertebrates like us are blessed with a second, more sophisticated set of immune functions. This additional line of defense is called "adaptive immunity," and it is indeed impressive. It allows the organism to target, seek out, and destroy specific invaders. And it has the power to "remember" the body's history of invasion. So this second kind of defense can quickly regain focus on specific pathogens that have been fought off on some past occasion. The immune system doesn't have a brain. Yet it's hard to resist applying labels like "smart" to the adaptive immune system. That's because it can (in its own way) "learn" and "respond intelligently" to pathogens.

When the adaptive immune system gets activated, we call it an immune response. Anything that can provoke an immune response we call an antigen (meaning something that can cause a defensive reaction). An antigen is some part of an invader – maybe a toxin, an enzyme, or a characteristic protein – that can be used to reliably identify it. Adaptive immunity directs its response specifically against the antigen that triggered it.

This specificity is one of two things that distinguish adaptive immunity from innate immunity. The other thing is memory. After a first encounter with an antigen, the antigen is remembered. Later, when that antigen is detected again, adaptive immunity already "has its number." The immune system already knows what to do. It has a running start enabling it to mount a quicker, stronger, and more effective response. The second time around the same invader often doesn't stand a chance against the rapid, informed immune response.

The way adaptive immunity actually works is amazing. Though we say the immune system "recognizes" an invader, that's not accurate. It can't recognize the way we can. Rather, some very limited feature of the invader – the shape of one part of a characteristic molecule, for example – happens to fit a particular antibody exactly. And that triggers an immune response against that particular shape.

To put this in human terms, it's as if we were blind, but had memorized the shape of our friend Gloria's nose exactly. Then if in our blind groping we happened to grab and recognize that nose shape, we would conclude (logically enough) that Gloria was there. We could call this "recognition-by-trigger." Evolution has chosen this as a practical solution for mounting an immune response. And it works really well.[53]

How adaptive immunity works

White blood cells called lymphocytes are the main actors in adaptive immunity ("lympho-" from the Latin "lympha" meaning water; "cyte" from the Greek "kytos" meaning hollow vessel). There are many different kinds of lymphocytes, but we'll focus on just a few here.[54]

The surface of some lymphocytes is covered with thousands of detectors. The detectors on any one cell are all tuned to "grip" or lock on to the same specific antigen. This enables that lymphocyte to "feel around" for the characteristic shape of that particular antigen, and to recognize it on contact. When it does find and lock onto this shape, the lymphocyte starts multiplying like crazy. The result is a horde of new lymphocytes, all identical to the first.[55] Each of these new cells is equipped with these duplicate detectors. The antibodies are ganging up on the antigen.

Each of the activated lymphocytes begins emitting streams of antibodies capable of reacting to that specific antigen. The antibodies circulate throughout the system. When they bump into their antigen, they stick to it. This both cripples the foreigner and marks it as a target. Other components of the adaptive immune system (for example the macrophages) will then be able to recognize and destroy the invader.[56]

Some lymphocytes develop in the thymus gland, and so are called "T-cells." Others mature in bone marrow, and are called

"B-cells." When a lymphocyte detects an intruder, it rapidly reproduces. It and its offspring begin producing antibodies specific to the foreigner that it has detected.

How do lymphocytes know how to detect intruders? Well, they don't "know." The immune response is not guided by intelligence or instructions. Our bodies have the ability to create billions of different antibodies with random shapes, and are doing so constantly. An enormous number of different lymphocytes are roaming the body, representing countless different molecular shapes, each capable of matching the antigen on a different sort of invader.

Only one out of a million lymphocytes may initially be able (by chance) to react to a particular invader. At first only this Lone Ranger is capable of producing the correct antibodies. But soon this one cell has copied itself billions of times. And each one of these copies is then able to produce the same antibody – the one specific to the foreigner. This selective process explains how an antibody precisely tailored to match an invader can be mass-produced by the immune system.

The Self / Other distinction

This system works only because it is able to distinguish chemicals that are parts of the body from chemicals that are foreign.

Elements of the immune system are engines of destruction. That's their whole purpose. Lymphocytes are invaluable in protecting against attackers. But they can do enormous damage if they attack the body itself. So it's absolutely essential that lymphocytes be able to tell the difference between invaders and components of the body. If your lymphocytes have trouble telling the difference you end up with autoimmune (immunity against the self) disorders, for example rheumatoid arthritis or lupus.

But the ability to make this distinction isn't easy to come by. There are thousands of complex chemical and biological products that are parts of the "self." Lymphocytes must not damage any of these.

The cells of the immune system – the lymphocytes, macrophages and others – must learn to tolerate every tissue, every cell, every protein in the body. They must be able to distinguish the hemoglobin found in blood from the insulin secreted by the pancreas from the vitreous humor contained in the eye from everything else. They must manage to repel innumerable different kinds of invading organisms and yet not attack the body... The immune system must be able to recognize foreign products whose chemistry is only slightly different from its own molecules.[57]

As if things weren't complicated enough, the forces of evolution are prevented from "hard-wiring" into the genes the ability to recognize the self. That's because every person is a different biochemical self.

The specific combination of chemicals that make each one of us up is unique! It's not possible to identify the components of an individual's body before that individual has been conceived. Each person has his or her own special mix of bodily elements. Why? Sexual reproduction ensures that we are all different by mixing genes from male and female. That's why an organ transplanted from my body will be rejected by yours. My tissues are so different that your immune system interprets transplanted tissue as an invader and attacks it.

Somehow your own adaptive immune system must unerringly recognize your individual chemical "self" – even though there are as many selves as there are people. Distinguishing reliably between all of those different chemicals is a job that might be tough for a supercomputer. How can mere cells, with no brain, make such a sophisticated discrimination? The fascinating answer only recently became clear.

A researcher named Lederberg came up with a mechanism that explains the self/other discrimination. He speculated that a great many more lymphocytes begin life than finally end up "on the job." During their development, a selective process takes place. Those cells that (by chance) react to components of the body are culled out and eliminated (killed by the body). This creates a "hole" in the developing collection of lymphocytes that exactly matches the set of chemicals making up that individual body.[58] The remaining lymphocytes – those that have not reacted to the body – are the only cells put into service in the immune system. Lederberg's scheme is called clonal deletion theory.[59]

This ability to self-discriminate is important enough that the body has evolved additional ways to destroy all lymphocytes that can attack the self. For example, a mature lymphocyte will usually be killed if it responds to a self-product without receiving a second chemical confirmation message.[60]

This is a neat system, and works very well most of the time. But there is a problem. What if a virus or bacterium is present in the body while the system is "learning?" Then the components of that invader will be accepted as "self." From that point forward the invader can never be recognized as something to fight![61]

Tricking the immune system in this way is not just a theoretical possibility. It happens. Researcher W. Paul describes an experiment showing that infection with a virus at an early age causes the organism to accept the virus as self. This particular experiment used lymphocytic choriomeningitis virus (LCMV). When newborn mice are inoculated with LCMV, the infection spreads rapidly without symptoms throughout their tissues. This happens because their immune systems – immature and not yet functional – learn to tolerate the LCMV antigens as harmless components of the self. From this point forward their immune system ignores the virus.[62]

Once a cell is infected by a virus, there is no way to clean the virus out of the cell. Instead, lymphocytes direct macrophages to destroy the infected cells. This stops the production of viruses by the cell machinery. This strategy is like burning down a barn to get rid of rats, but there's no other way to stop the spread of the virus.[63] It's easy to see that if a virus were to spread too quickly, the immune system's attempts to contain it might do more harm than good. The "barn-burning" might leave path of destruction in the wake of the spreading virus, but never quite catch up to it.[64]

When this happens the damage wreaked by the immune response could outweigh the damage done by the virus. As if realizing the futility of the situation, the immune system may at some point relent. As Paul notes, "the immune system may sometimes subvert its own reaction to viral infections if that response would hurt the host more than the pathogen."[65] In such cases all lymphocytes capable of recognizing the invading virus are destroyed or otherwise inactivated.

This way of reining in the immune system can have serious consequences. The elimination of those antigen-specific T cells leaves a hole in the immunologic defense. It will never again be able to fight that particular virus.[66]

Viral inhibition of immune response

The body's struggle to cleanse itself of viruses is unceasing. The conflict between virus and immune system is an ongoing war. Viruses continually innovate to stay beyond the reach of the immune response. In order to survive, all viruses develop ways of subverting and evading immune defenses. They all attempt to inhibit immunity in one way or another.

Viruses use a variety of strategies to thwart immunity.[67] [68] A good number of these involve what could be called "trickiness" or "stealth." Viruses are simple pathogens, certainly not complex enough to act intelligently. In fact there is debate about whether they are even alive. Yet the strategies they employ do seem crafty and insidious. The Human Immunodeficiency Virus (HIV) provides some great examples.

Most viruses reproduce within tissue cells. The HIV virus directly undermines the immune response by infecting the T-lymphocytes (primary agents of the immune response). They reproduce within them, ultimately destroying them one by one. The ongoing decimation of the lymphocyte population gradually narrows the range of antibodies the lymphocytes can produce. So this strategy progressively weakens host defenses. As variability among the lymphocytes declines, the likelihood of a lymphocyte matching an invader becomes smaller and smaller. So resistance to disease in general declines. This is the reason for the appearance of opportunistic infections in later stages of HIV infection.

The HIV virus particle evades immune response in a second way. It dons a "disguise" to avoid detection. A flow of new virus particles is produced by cell machinery within infected lymphocytes. Once the virus particles are assembled, they pop out through the cell wall of the lymphocyte, on their way to infect other lymphocytes. But in the process they remove a bit of the wall material from the host lymphocyte and wrap themselves in it. Any lymphocyte later encountering the wrapped-up virus particle is fooled into treating it as part of the body itself. Why? Because the selective process through which lymphocytes learn to distinguish self from other guarantees that they will not be able to react to the self material surrounding the camouflaged virus particle. HIV is in this sense a wolf in sheep's clothing.

HIV possesses yet another trick, one characteristic of many viruses. HIV continually changes the shape of its component molecules through mutation. By the time the lymphocytes are producing antibodies to a particular antigen molecule, later generations of viruses are likely to have developed a molecule with a different shape. The antibodies no longer fit the mutated molecule, and so cannot bind to their intended targets. So the shape-shifting virus is constantly slipping away from the immune system.

All of these measures help to overcome the host's immune functions. This kind of "sneakiness" is characteristic not just of viruses, but of all types of parasites. That's because any parasite that isn't able to sneak around immune defenses is a dead parasite. By destroying all the incompetent parasites, the host's immune system creates very strong selection pressure to make them sneakier and sneakier over time. Conversely, parasites' tricks becoming ever more effective creates strong selection pressure upon the host. This pressure pushes development of the immune system to ever more sublime levels. The result is an arms race that never ends. And the products of that arms race are some of the most mysterious and magnificent characteristics of organisms.

Screw it up or die

The thing to remember about all this is that viruses always mess with the immune system. Logically we can see why this has to be. Every parasite that can be detected, however weakly, by immune defenses comes under selection pressure to counteract those defenses. Viruses have evolved a number of specific strategies of evasion. Some "hide" in tissues that are out of reach of the immune system.[69] Others produce special chemicals that interfere with the host's normal immune response.[70] Researchers Borrow et al note that though the entire range of mechanisms by which viruses mess up immunity have yet to be fully clarified, the reasons why they do so are clear. They do it to avoid being eliminated. But messing with immunity globally may significantly weaken the host.[71]

Virus infection is frequently associated with a transient or more long-lasting generalized suppression of the host immune response. While the induction of immune suppression is potentially advantageous for the virus, being one of the mechanisms by which viruses may escape clearance and establish a persistent infection in vivo... the clinical consequence for the host may be severe.[72]

Button pushers' paradise

How do simple things like viruses gain the complex abilities they seem to have? There's a simple answer, and it's this: They randomly "push buttons" within the host until they stumble upon something that promotes their survival.

Living things need a set of abilities to stay alive. Over evolutionary time they develop the capacity to make things happen. Cells' abilities are based largely in the biochemical action of proteins, chemicals of varying shapes that let them perform functions essential to life. The abilities of a cell are stored in the form of DNA. The DNA strands serve as templates to create the cell's various proteins. What we call genes are specific DNA sequences. Genes are activated on demand, and translated to proteins to meet the organism's current needs.[73]

The greater the number of genes, the greater the number of unique proteins the organism can create. Each protein has a shape that confers some specific ability – the shape does something. The greater the number of proteins, the more things that organism is able to do. Complex organisms could only come into being as cells developed the ability to create and store within themselves a large number of genes. Just as important – they had to have created tools to manage those genes.

As organisms evolve toward complexity, they accumulate more and more capabilities to aid their survival. That's good for the organism. But it's also a gift to the organism's parasites. As the number of host genes grows, parasites have more and more genes to exploit. We can build a helpful analogy by calling those host genes "buttons." Each new host capability provides another button for its parasites to press.

By design, parasites are less complex than their hosts. Parasites are simpler, and because they are simpler they can evolve faster. The strategy of parasites is to run "lean and mean." It's very, very difficult for any host organism (burdened as it is by big suitcases stuffed full of genes) to evolve fast enough to outrun its parasites (which travel light). Parasites don't pack toothbrushes or clean shirts to take with them on their evolutionary journey. They don't have to. They will steal most of what they need from the host.

Every ability that belongs to the host is potentially part of the parasite's extended phenotype. The host is limited to the set of abilities it has accumulated for itself over evolutionary time. In order to show complex behavior, the host needs to have accumulated a complex set of capacities. It also needs to have integrated those capacities so that they work together harmoniously.

But the parasite doesn't have to jump such hurdles. It doesn't have to be complex to show complex behavior. It just has to manipulate its host into doing something complex. And that means simply pushing the right buttons. We could say that all the talents of the host are at the parasite's disposal.

In days of old, computer programmers had to figure out exactly how to get their machines to do what they wanted. They had to build into their programs detailed routines to perform each desired action. That was tedious. But in our modern era virtually all computer languages provide a rich set of procedures and functions that do anything you want. They're built right into the language. Now it's just a matter of deciding which functions to call, and learning the syntax to call them properly. Programmers don't have to know how these functions actually work. In fact the inner workings of the functions are deliberately hidden from the programmer.

These days a well-educated programmer is like a graduate of Hogwart's School of Witchcraft and Wizardry. Once students learn to cast the spells properly, they can cause magical things to happen. They're magical because programmers can make their computers perform wonderfully complex actions – even though the programmers remain essentially clueless. They don't know (*can't* know, really) how they're doing what they do. They simply learn to call the right procedures in the right way, to "utter the correct incantation."

Parasites use a similar approach to work their brand of magic. It's as if the parasite were sitting at a panel with rows and rows of buttons, each of which does something. Having no brain, the parasite doesn't know what button to press. It doesn't have to. Neither does it have to comprehend what the buttons actually do. Its strategy is much more basic: to press buttons until by chance it hits one that helps it survive. Those individual parasites whose mutations happen to let them press the right buttons do survive. Then they pass their successes on to their future generations. The ones that guess wrong take their errors to the grave with them. So over time parasitic "wisdom" (again the quotes) is accumulated and a "strategy" unfolds.

The common cold is caused by a viral parasite – a rhinovirus ("rhino-" from a Greek word meaning "nose). Its strategy is to multiply rapidly within its human host. It uses its short time aboard to reproduce and transmit itself to a new host. The host immune system typically mounts an effective response within a week or two. By the time the virus has completely used up its welcome, it has already moved on to a new host. So although the virus doesn't have a purpose (or even a brain), it is operating with a strategy. And messing with your nose is part of that strategy. The virus causes irritation, nasal discharge, and sneezing. The discharge and sneezing sprays virus particles all over, transmitting the viral pattern to new hosts. Shedding of viruses may start a few days before cold symptoms have become obvious. Shedding reaches a peak on days 2-7 of the infection, which may last as long as 3-4 weeks.[74]

This strategy is not only clever, but it involves some complicated host behavior. Did the virus plan the strategy? No. The virus can't plan. Does the virus itself have the right DNA to produce the right proteins to precisely control mucous production or the behavioral sequence of a sneeze? No. It causes these effects simply by "pressing buttons" within the host. That's virus magic.

It only looks complex

Parasites can enhance their survival by provoking organisms to perform complex actions – behaviors those organisms are already capable of performing. The capacity for a coordinated sneeze comes from the human being – from the host, not from the viral parasite. The readiness to perform a sophisticated action is part of the host's normal behavioral repertoire, and exists before the parasite pays its visit. The rhinovirus merely triggers it.

We must assume that the button-pushing elements of the rhinovirus' transmission strategy came together gradually over generations under the influence of evolutionary forces. So although the parasite is a simple creature, its influence is manifested in complex behavior. Our sneezes and our runny noses truly do belong to us. But they are also part of the extended phenotype of the virus, when that virus is inside us. The lesson here is that it's possible for simple parasites to evoke complex host behavior.[75]

In summary

Parasites are everywhere in the animal kingdom. They survive by hijacking resources from their host organisms. This places their hosts at a relative disadvantage in their struggle for survival. So the host species is under strong pressure to develop effective means of coping with them.

This selection pressure has over time led to evolution of a sophisticated set of coping mechanisms in the host. We classify many of these under the heading of immunity. Simple organisms have a basic set of mechanisms biologists call innate immunity. This set sensitizes organisms to the presence of elements that "don't belong" in the organism, and enables a generic sort of defense against these foreigners. Vertebrates are more sophisticated. They have, in addition to innate immunity, a set of responses called adaptive immunity. These allow the organism to recognize specific intruders, to target them individually and specifically, and to remember them in a way that makes for rapid response in the future.

For their part, the parasites are obligated to find ways to defeat host immunity. They deal with it or die. This makes for intense selection pressure. Over vast periods of time, and under this pressure, parasites have evolved multiple "sneaky" ways of evading detection by disabling or overcoming host immune response.

The host has no choice but to deal with the parasite. So it develops immunity. The parasite has no choice but to thwart that immunity. So it develops evasive strategies. The result is a never-ending arms race.

The arms race has led to the evolution of some of the more mysterious features of life, features that in the past have confused scientists. Immunity as resistance we can easily understand. In the next couple of chapters we will look at two more mysterious features. These are aspects of our humanity that in the past have been particularly difficult to fathom – sex and death. They may not at first seem related to immunity. But now we're ready to view them in a new light.

Chapter Three
Sex and parasites

A nimals' need to protect themselves from parasites (including viruses) has led to some of the most dramatic of biological adaptations, the most fantastic of evolutionary developments. Among these adaptations is sex, a critical aspect of our human existence that had been utterly mysterious until recently. Sex has been newly interpreted as a parasitic defense. This chapter explains how sexual reproduction came to be, and how it works to defend species.[76]

The rise of the eukaryotes

Sex. Hmmm. This is quite a story, and one with lots of convolutions. Maybe it is best told by beginning at its beginning. It begins with an appreciation of the inner workings of cells.

Our human bodies are made up of complex cells that biologists call "eukaryotic." That name comes from Greek words meaning "good kernel." That most excellent kernel is the cell nucleus. It's a membrane-enclosed structure that houses our DNA. Most eukaryotic cells also contain other membrane-bound organelles – for example mitochondria. All complex multicellular organisms are made up of eukaryotic cells – animals, plants, and fungi.

The earth is about four and one half billion years old, and life had its start a little less than four billion years ago. Eukaryotic cells, with all their complexity, seem to have appeared relatively recently in the history of life – only a couple of billion or so years ago (there is debate about exactly when).[77] Preceding them by one or two billion years were two other less complex cell families termed "prokaryotes." These families, the bacteria and the archaea, generally exist as one-celled animals. They are one-celled because prokaryotes don't have the capabilities to meet the complex requirements of multicelled organisms.

In most prokaryotic cells the DNA sits naked in the cell cytoplasm, the jelly-like stuff inside the cell membrane.[78] There's no nuclear membrane enclosing the DNA. Prokaryotic cells also lack other internal membrane-bound structures. Compared to eukaryotes they look primitive. Having originated about 3.5 billion years ago, bacteria are thought to have been the first membrane-enclosed life forms to evolve on earth. Archaea apparently arose not too long (in geologic time) after that. But it took another one or two billion years for the simplest eukaryotic cells to make their appearance. The fossil record says complicated multicellular lifeforms (which are all made up of eukaryotic cells) first appeared in the Ediacaran period "only" 635 million years ago.

The Ediacaran era was named after the Ediacara Hills of South Australia, where fossils of the creatures typical of that period were first discovered. It seems that by this time one-celled eukaryotic animals had already been around for awhile, and that some multicellular life forms were beginning to take shape. Even so, during this period the multicellular organisms seem at first to have been relatively simple soft-bodied creatures. They lacked the "biomineralized" hard parts that could be easily preserved as fossils. So the fossil record from that period isn't very clear.

The Cambrian explosion

About 530 million years ago everything changed. The fossil record says the simple Ediacaran-type animals were somehow swept away by a brand-new lineup of exceptionally varied animals. This happened during a period of rapid development called the Cambrian Explosion. All the complex lifeforms we know today have their roots in this relatively brief (geologically speaking) period. We can only conclude that this period marks some unparalleled changes in the nature of life on this earth. As paleontologist Ben Waggonner puts it,

> *[T]he "Cambrian explosion" was not only a major evolutionary radiation of taxa [groupings of animals]; it encompassed the widespread adoption of ecological strategies and modes of interaction that were rare—though not entirely absent—in Ediacaran biotas [the taxa of the edicaran period]. Furthermore, the "Cambrian explosion" may not have encompassed the origin of the animal phyla [the major groups of animals]... but it must have encompassed the appearance of evolutionary innovations that made these new strategies possible.[79]*

What happened back then? The exact reasons for the Cambrian blooming of life are still being debated. But the Explosion certainly had to do with the growing complexity of eukaryotic cells. This new complexity empowered them to support cellular differentiation and specialization.

Even before eukaryotes came onto the scene, some prokaryotic cells had developed the ability to assemble themselves into very simple multicellular organisms. These weren't anywhere near as fancy as the kind of multicelled animals we know today. Multicelled prokaryotic organisms took the form of uncomplicated "filaments, clusters, balls, or sheets of cells that arise via a mitotic division from a single progenitor."[80] The only cell differentiation that was common was that separating cells of the body and cells reserved for reproduction.

These prokaryotic multicelled animals were pretty basic. They were assembled out of nearly identical units just stuck together. The organism didn't have any internal organs to pass nutrients around, and it didn't have any way to regulate operations inside it. Every cell in the organism needed to be in direct contact with the water, so it could absorb oxygen and nutrients. Complex three-dimensional structures (organs, circulatory system, etc) were completely out of the question – they just weren't possible given the limited capabilities of prokaryotes.[81] Three-dimensional forms such as internal organs had to await the arrival of the eukaryotes, with their much greater capacity to support cellular differentiation.

What is cellular differentiation exactly? During a multicelled animal's development, "generic" cells called stem cells change into a variety of different cell types. Afterwards each type is able to perform specialized functions for the multicelled organism as a whole. Once specialized through differentiation, cells cannot return to their earlier state. They are "locked in" to their assigned jobs.

It seems likely that what triggered the Cambrian Explosion was eukaryotes' new ability to assign cells specific and permanent jobs. This capacity is typical of the highly developed animals that we know today (including us humans). In the multicellular organisms before this time the prokaryotic cells could manage some limited specialization of function. But even with their specializations they weren't permanently committed to their specific roles within organisms. So they weren't truly differentiated. Observer Chris Phoenix speculates,

> *Although several factors have been identified that may have enabled the Cambrian Explosion, there remains the question of the final trigger. The development of cellular differentiation, the capacity of a cellular lineage to become specialized permanently in response to external signals, is proposed as the trigger. Cnidaria [the group of animals that includes the jellyfish] and eukaryotic slime molds, which pre-date the Cambrian, have specialized cellular functions but their cells do not appear to differentiate irreversibly. With differentiation, cellular roles can be fixed during embryogenesis [the development of the embryo]. Fixed cellular roles may have enabled more complex and more competitive organisms, perhaps by facilitating multi-scale structures in organs or increasing the stability and reliability of network systems including synaptic connections.[82]*

With the coming of cellular differentiation organisms gained the power to operate something like societies. Each cell type could do highly specialized work in support of a larger organism.

For the greater good

The new capacity for differentiation led to complexity that hadn't been possible in any previous era. Within the new Cambrian creatures individual cells could act as if they were a variety of very different one-celled creatures, all working cooperatively to support the larger multicelled organism. This gave multicellular animals dramatically increased capabilities. The ability to differentiate improved survival.

Though the cell types making up a eukaryotic multicelled animal look and act very different, each contains exactly the same package of DNA. This package can be passed on to future generations. Because they all share the same DNA, the fates of all the individual cell types are tied to the fate of the animal as a whole. For their survival cells no longer had to depend upon the fitness they exhibited as individuals. An entire "team" of very different cell types could pass on a single DNA package – one representing the whole team – to later generations. Suddenly, body cells were expendable, because their individual deaths did not stop their DNA from being passed on to the next generation. In fact, the death of some cells was necessary to make multicellular life possible. But more about that later.

The exotic eukaryotic

The differences between prokaryotic cells and eukaryotic cells are bewildering. Prokaryotic cells are small in size, and simple in their life strategy. They don't carry very many genes. They are committed to making do with very little, and reproducing as quickly as possible. They don't have any specialized organs for respiration, so they must rely on their external cell membrane to "breathe." They can't increase their tiny size too much. If they get too large, the ratio of their surface area to their volume becomes too small to support respiration, and they suffocate. They don't have a system for handling a big set of genes. So the pressures of natural selection relentlessly push prokaryotes toward smaller physical size, as well as smaller gene store.

The way prokaryotes store and access their genes is crude in comparison to the methods used by eukaryotes. Prokaryotes use a method something like the tape drives in early computers. Their DNA is stored in a simple loop. It's accessed by reading sequentially around the loop. As a result they can't afford to carry too many genes – doing that would overload this simple storage and retrieval system. A large number of genes would overwhelm their mechanisms for protein synthesis and reproduction.

Because of these limitations natural selection pushes prokaryotes to keep their gene stock small and manageable. There is evidence that bacteria even find ways to get rid of parts of their DNA that they aren't using at the moment. As writer Nick Lane explains,

> *Bacteria are pared down to a minimum compatible with a free-living lifestyle. They are ruthlessly streamlined, everything geared for fast replication. Many keep as few genes as they can get away with; they have a propensity to pick up extra genes from other bacteria when stressed, bolstering their genetic resources, and then lose them again at the first opportunity.* [83]

Their modest store of genes limits prokaryotes' capabilities. Their proteins restrict them to a small number of activities. So they just can't handle the complexity it would take to differentiate into multiple cell types, a minimum requirement for multicellular organisms. The enduring success of prokaryotic cells over the ages is due largely to this fact: They are simple forms, "hardwired" through selection pressures to do a single job very efficiently. The way they have accomplished this job has made them quite successful. But it has for billions of years kept them in an evolutionary rut. "Nothing is more conservative than a bacterium," opines Lane.[84]

We should not mistake bacterial conservatism for inferiority. Bacteria are good at what they do. So good that complex animals today harbor them to perform vital life functions for them. We humans, for example, could not survive without the chemical products of the trillions of bacteria that live inside us. The human genome can code for about 22,000 proteins. But the bacteria within us contain genes coding for about 8 million unique proteins. The humbling truth is that "our" bodies contain ten times more bacterial cells than human cells![85] Since they can outvote us, maybe our bodies really belong to our bacterial partners.

Marvelous multicellularity

Eukaryotic cells are quite different from prokaryotic cells. For starters, they are generally bigger – much bigger. Their volume is typically hundreds of times that of the average prokaryote. Critically, they have a much more sophisticated system for storing and retrieving genes. Consequently they can handle a gene database many, many times larger than prokaryotes. Their ability to store more genes reduces the selection pressure to jettison the ones they aren't using at the moment. So eukaryotes do store many times more genes than prokaryotes.

Their larger and more diversified gene store means eukaryotes can make lots of different proteins. That translates to doing lots of different things. The greater range of capability is what enables cellular differentiation. Eukaryotic stem cells are transformed into a large number of distinct cell types. The different types perform the diverse types of work required for the complex life of a multicellular organism. So all true multicellular animals are built of eukaryotic cells.

Just look at human bodies. Our bodies are an assembly of eukaryotic cells. These are differentiated into a large number of cell types – muscle cells, liver cells, blood cells, and neurons. The DNA information in each of these cells is exactly the same. But from their appearance and activity alone you might think that each type was an entirely different organism. How can they look and act so different?

Within the human body each cell type plays a specialized role in maintaining our total life functions. These specialists can do this because each cell type expresses a different subset of the animal's total DNA library.[86]

A well-equipped kitchen has a variety of ingredients and cooking tools. But by selecting different subsets of the ingredients and the tools, the cook can prepare a large number of very different dishes. Likewise each cell type looks and acts so different because it is using different selections of its total DNA. So in terms of gene expression, protein production, and structure they really do behave as if they were different organisms.

The value of emulation

A computer can be set up to behave as if it were another completely different kind of computer, a different "virtual machine." We say that one computer is "emulating" another. Although Computer One (maybe a Mac) is really not at all like Computer Two (maybe a PC), the emulation makes it act just as if it were. So programs designed for Computer One can run on Computer Two.

Through differentiation, a eukaryotic cell can emulate many different kinds of one-celled organisms. Maybe we could call these "virtual organisms." A nerve cell and a liver cell aren't different organisms – but they certainly look and act as if they were. That's because they are using different parts of the same human genome. This capacity for emulation allows each of the different eukaryotic cell types to perform highly specialized functions. The different cell types, each doing a different job, evolved to work together. They act harmoniously to fill the complex needs of the larger multicellular animal that is the human being. If eukaryotic cells could not express gene subsets, complex multicellular organisms simply could not exist.[87]

A cellular mystery

Eukaryotic cells were the first cells on earth with the complexity needed to emulate a variety of different cell types. There is general agreement that their arrival was a spectacularly significant development in the history of evolution.

But how did eukaryotes get here? Their features are so vastly different from the simpler bacteria and archaea that their origin has long been a source of puzzlement and debate among paleobiologists.[88] [89] Among the mysteries presented by eukaryotic cells are these:

* Unlike prokaryotes, eukaryotes have a nucleus that segregates their DNA – and the DNA transcription process – from other cell structures and processes.

* Eukaryotes are chock full of all kinds of complex and specialized structures: for example nucleus, golgi complexes, membranes, chloroplasts (in plants), and mitochondria.

* Eukaryotic animals generally reproduce sexually, a means completely unlike the simple splitting that prokaryotes use to reproduce.

* Eukaryotic cells have a complex internal "skeleton" made up of microtubules. They have the ability to quickly alter

the specific shape of this skeleton. This permits them to move in ways prokaryotes can't manage.

* They store their DNA using a method quite different from the method used by bacteria or archaea. As noted earlier, bacteria store DNA in a loop, analogous to the tape drive of an earlier computer era. Eukaryotic cells maintain their DNA in straight segments organized into a number of chromosomes. This is a system more like the randomly accessible data on the hard disk drives of present-era computers.

* The genes in eukaryotic chromosomes are broken up into subunits called introns that are separated by non-coding segments called exons.

* Eukaryotic cells translate their genes into proteins in a mystifyingly complex manner that involves translating DNA into RNA within the nucleus, then snipping out the exons (and sometimes rearranging the introns) before joining the introns together into a final RNA product.

* The RNA is transported out of the cell's nucleus before being translated into proteins.

* The size. Eukaryotic cells are (relative to the size of their ancestors) huge.

* Eukaryotes are capable of endocytosis. They can take in particles or molecules from outside the cell by enclosing them in a bubble or vesicle pinched off from the cell membrane. Prokaryotes are as a rule unable to do this.

These differences are profound. Here's something that boggles the minds of paleobiologists: It isn't clear where all these changes came from. Biologists conceive of natural selection as taking place in small steps. Perhaps these steps are recorded somewhere in an undiscovered part of the fossil record. But fossil evidence of cellular evolution is very scanty.

We do have non-fossil evidence we can use to resolve part of the mystery. The information contained within modern-day gene sequences reflects the ancient origins of organisms. Examining this information, we can draw some conclusions. Eukaryotes contain a combination of genes from the two different types of prokaryotic cells – archaea and bacteria. So the eukaryotic cell appears to have sprung into existence as an unlikely mashup of the genes and structures of archaean and bacterial cells. As Lane notes,

> *The more genes that we study – and one recent analysis combined 5700 genes, drawn from 165 different species into a "super tree" – the more plain it becomes that the eukaryotic cell did not evolve in a standard "Darwinian" way, but rather by some sort of mammoth gene fusion. From a genetic point of view, the first eukaryote was a chimera – half archaea, half bacteria.*[90]

Genetic information can be used to draw tentative pictures of the eukaryotic family tree. Researchers Kelly et al performed a high level analysis of genetic information contained within sequenced genomes of archaea, bacteria, and eukaryotes. Their analysis says an ancient, simple ancestral cell line split into archaea and bacteria a very long while ago. Following their parting of the ways, archaea and bacteria each continued to diverge into multiple species. The eukaryotes came into being sometime later when an organism from an archaean line somehow ended up with a bacterium living inside it. Evidence summarized by Kelly et al shows that this coming together happened a long time after archaea and bacteria split off from one another.[91] Amazingly, this seminal event appears to have occurred only once. We know this because all existing plants and animals are so closely related (biochemically and statistically speaking) that they all have to have arisen from exactly the same source.[92]

So a really important event in the eukaryotic lineage was marked when a bacterial cell – the proto-mitochondrion – somehow started living inside an archaean cell. That bacterium would have been the first primitive mitochondrion, and the archaea/bacterium combo would have been the first eukaryote. How did this union come to be? There's no way to be absolutely sure.[93] But researchers Koonin and Aravind did a careful review of data on genome sequences. They and others think the bacterium may have originally been a parasite living inside an archaean host.[94] [95]

Endocytosis and more

Reading through the list of prokaryote/eukaryote differences above, we could easily judge the last one – endocytosis – to be among the least significant. But there is evidence that endocytosis was the first in a chain of falling dominoes that led to modern eukaryotes.

Endocytosis (from the Greek "endon," meaning within, and "kytos," meaning hollow vessel) is the ability of a cell to draw things from outside it, and put them in packages inside. The eukaryotic cell draws things through its covering – the plasma membrane – and in the process bundles them up into neat packages. Endocytosis comes in multiple flavors. For example eukaryotic cells can "sip" liquids from outside (pinocytosis); they can insert molecules into standardized containers having a buckyball-like skeleton (clathrin-mediated endocytosis); and they can "eat" objects (phagocytosis) – even large objects like other cells.[96]

At one time scientists viewed endocytosis as a relatively simple process by which the cell transported molecules through its outer covering, the plasma membrane. But recent research has shown that endocytosis occupies a central place in the design and functioning of the eukaryotic cell itself. Prokaryotic cells had used relatively unsophisticated methods, for example channels or pumps, to draw sugars, amino acids, and ions through the cell membrane. But according to researchers Sigismund at al, as these cells evolved these "entry portals" grew more complex. This could have happened as the oceans grew larger, and nutrients in the water around the cells grew more dilute.[97]

Once the capacity for endocytosis was there, the cell began to adapt it for other uses too. The operations that make cellular differentiation and specialization possible are complex. Eukaryotes could not have evolved without a sophisticated program of cellular signaling – a system telling the cell what to do, where to do it, what chemicals to do it with, and exactly when it should be done. In modern eukaryotic cells, say these researchers, endocytosis has morphed into that messaging system. Over the course of time, endocytosis was transformed, as Sigismund et al put it, "from its primordial trade into something rather different: a powerful communication and compartmentalization infrastructure or, in essence, what we define as 'the cell logistics of the cell.' "[98] In modern eukaryotes endocytosis is "inextricably linked with almost all aspects of cellular signaling."[99]

> *Endocytosis is actually the master organizer of cellular signaling, providing the cell with understandable messages that have been resolved in space and time. In essence, endocytosis provides the communications and supply routes (the logistics) of the cell.... Endocytosis constitutes one of the major enabling conditions that in the history of life permitted the development of a higher level of organization, leading to the actuation of the eukaryotic cell plan.[100]*

So it seems likely that many of the sophisticated capabilities of the eukaryotic cell are derived from a single evolutionary development. Much of the good stuff that makes cell differentiation and specialization possible started with endocytosis. This good stuff includes the nuclear membrane, the complex system of internal membranes, the golgi apparatus, intercellular (between cells) communication via exocytosis and neurotransmitters, and multiple components of the immune system.

It appears the development of endocytosis may have been the

first step along a path that included evolution of complex internal structures and a signaling system. That in turn made for more precise control over cell operations. Better control, coupled with development of the mitochondrial energy source (which permitted greater cell size), paved the way for the larger DNA stores required for cellular differentiation and specialization.

Uninvited guests

All the increased capabilities deriving from endocytosis were tremendously helpful for evolving complex life upon the earth. Unfortunately these improvements were available not just to the cells that developed them, but to the viruses that would use them as hosts. As the capabilities of eukaryotic cells multiplied, the intricacies of viral exploitation easily kept pace. And viruses did indeed exploit host endocytosis.

Of the viruses that now infect eukaryotic cells, "most depend on endocytic uptake, vesicular transport through the cytoplasm, and delivery to endosomes and other intracellular organelles."[101] We know that today viruses invading modern eukaryotes take full advantage of their hosts' encapsulation and transport mechanisms.

> *Endocytic vesicles ferry incoming viruses from the periphery to the perinuclear area of the host cell, where conditions for infection are favorable and distance to the nucleus minimal. This allows viruses to bypass obstacles associated with cytoplasmic crowding and the meshwork of microfilaments in the cortex.*[102]

A bonus for viruses is this: when they sneak in using endocytosis they can "hide" within the resulting vesicles to avoid detection by the host immune system. All in all, the endocytic path into the cell provides so many advantages that even those viruses capable of penetrating directly through the plasma membrane often use the endocytic route instead.[103]

Even with the burden of viral tag-alongs, it seems clear that the advanced capabilities of eukaryotic cells led to the rise of multicellular organisms. And multicellularity, in turn, led to the Cambrian Explosion, that dramatic blossoming of life a little over half a billion years ago.

Sexual awakening

The coming of multicellular life had an interesting side effect. With multicelled organisms, each cell's DNA is exactly the same as every other cell's. So a single DNA package represents the interests of every cell in the body. As a consequence, survival of the individual team members is no longer important. The reproductive interests of the individual cells making up the body are fully represented by the fitness of the complex larger organism. Now all that really matters is the survival of the larger organism. The team as a whole becomes the important thing, not the individual cells. Multicellular life created a new unit of natural selection!

This "selection by team" introduces a complication. When an organism is just a single cell, that cell can reproduce by fission, or "splitting." The cell represents itself. But when cells are differentiated into different morphological types, which of all the different types of cell represents the whole organism? Does it make sense to pass one cell type on to future generations?

The answer lies in the process of cellular differentiation. When a multicelled animal is first conceived, it starts off as a single undifferentiated cell. That cell begins to divide, and does divide many times. As it divides, the cells differentiate. They are "locked in" to specific types appropriate to their function in the body. Once differentiated, they cannot return to an undifferentiated state. But during differentiation a special set of cells – a set we call stem cells – is set aside. These are special "generic" cells that retain the potential to differentiate. These are the germ cells. They are the ones that will represent the entire team in the next generation.[104]

Most eukaryotic organisms don't reproduce by splitting. Eukaryotes developed an entirely new method of passing on their genes, a method that would have a dramatic impact upon the evolutionary process itself. We call this new method sex.

In sexual reproduction half the genes from a male are shuffled with half the genes from a female to produce new combinations of genes. It is the resulting mix of genes that passes on into the next generation. Now what sense can we make of this?

The mystery of sex

We're so familiar with sex that we tend to just accept it. In human terms it is identity, attraction, intense desire, deep pleasure, and a sense of fulfillment with a mate. But to scientists sex has always been something of a mystery. As a means of reproduction it seems (scientifically speaking) like a terrible idea. It's overly complex, inefficient, and more than a little clumsy. Worse, sex seems to run against what we assume is the whole point of natural selection – passing successful gene combinations on to the next generation.

Before sex was invented, cells represented themselves. The individuals most capable of making a living in their environment survived. These fitness champions passed on their winning formula to new generations as a complete package of genes. But with sexual reproduction all that changed.

Wrestling with sex

There's one particular thing about sex that's really hard to understand: When animals reproduce sexually they have zero hope of passing into the next generation as complete units. The genes of the fitness champions are always diluted 50-50 by mixing with the genes of another animal.

During sexual reproduction the genes of a male are randomly intermixed with the genes from a female. In the process novel combinations of gene alleles (variants of the same gene) are cobbled together – all of them drawn from the reproducing population's gene pool. The particular and unique gene patterns of the individual parents – winners of the previous generation's Darwinian fitness contest – are all mixed up. So sex throws away half a winning gene package, and scrambles up what's left. This seems like madness. At least, it seems like madness if you believe that the only purpose of reproduction is to pass on your genes.

In fact the rationale for sexual reproduction eluded biological theorists for quite a while. Nick Lane comments that "Some of the best minds in biology have wrestled with the problem of sex, but only an incautious minority have been inclined to speculate about its deep origin."[105]

How could scrambling up the genes help? Measured in terms of conservation of energy and other cellular resources, simple splitting of cells is much more efficient than sex. While the origins of sexual reproduction remain obscure, its popularity makes it clear that somehow, someway it must enhance survival. It simply has to be good for something, or it wouldn't be there.[106]

The lottery of life

So how can we make sense of sex? One possible rationale held more promise than most of the competing theories. In the early 1970's biological theorist George Williams proposed that having a set of genes is something like holding a lottery ticket.[107] If an animal's genes are a good match to its environment, then it wins the survival lottery. When a winning animal reproduces by splitting it's like bequeathing its descendants several identical lottery tickets, each with the number that won last time. But when a winning animal reproduces sexually it's like passing its descendants a handful of new tickets, each with a different number.

When does it make sense to hold a ticket with the number that won last time? When that number is likely to be called again. That's when an animal's environment is stable and predictable from generation to generation. So where the environment is stable we would predict that reproduction by splitting would most benefit the next generation.

On the other hand, where the environment is unstable and unpredictable it would make sense to get a bunch of new tickets, each with a different number. One of those new numbers might be a winner under the new set of conditions. So in an unstable environment sexual reproduction would be the best method.

That's the lottery theory. Could it be the explanation for sex?

Well, it certainly is logical. Just as important, it is testable. In the 1980's researcher Graham Bell did look at organisms in both stable and unstable environments. The lottery ticket theory predicts Bell would find sexual reproduction more frequently in unstable environments. But what he discovered was exactly the opposite, that "sex was most commonly practised in environments that are stable and not subject to sudden change."[108] In fact, the *most* changeable environments often harbored the *least* sexual species.

So the evidence is squarely against the lottery theory. Variable environments are not the key to understanding sexual reproduction.

Enter the Red Queen

Here's where things get really interesting. As it turns out, there is one particular set of circumstances where sexual reproduction does provide a clear advantage. What are those circumstances? The surprising answer is this: Sex helps when the environment is full of parasites. That's because the shuffling of genes through sexual reproduction makes it harder for parasites to maintain their grip on an organism.

The presence of parasites creates a very special selection pressure – one that seems almost paradoxical. Returning to our lottery ticket analogy, parasites ensure that the winning ticket will never be repeated. A winning number in the present generation is sure to be a losing number in the next.

Why is this? Because in the course of a generation parasites have time to adapt themselves to the winning DNA packages. The parasites will predictably break the codes of the current generation. Parasites convert winning numbers into losers. So if host animals pass their winning tickets to the next generation, they are setting their descendants up for slaughter.

Unwilling hosts

The "host organism" is euphemistically named. This host certainly does not voluntarily cater to its parasitic "guests." Recently gathered evidence suggests that the gene shuffling of sexual reproduction gives host organisms a way to "outrun" its parasites. More accurately, it keeps host and parasite running neck and neck.[109]

This is the "Red Queen" hypothesis, so named after the red queen in the book *Alice in Wonderland*. The queen tells Alice that in Wonderland people must run as fast as they can just to stay in one place. A fitting metaphor. If it is to flourish, a parasite would prefer that its host's genes stay exactly the same from generation to generation. After all, the parasite gains its powers simply by "pressing buttons." It doesn't know why a button works, or even what it does. Eukaryotes' sexual reshuffling of genes is the host's way of moving the buttons around, keeping the parasite guessing. The constant changes keep the host's DNA a little unfamiliar to its destructive parasites, so that the host is never completely overwhelmed by them.

The Red Queen hypothesis is eloquently summarized by writer Matt Ridley in his excellent book *The Red Queen: Sex and the Evolution of Human Nature.*

> *... Let us concentrate on viruses, bacteria, and fungi, the causes of most diseases. They specialize in breaking into cells – either to eat them, as fungi and bacteria do, or, like viruses, to subvert their genetic machinery for the purpose of making new viruses. Either way, they must get into cells. To do that they employ protein molecules that fit into other molecules on cell surfaces; in the jargon, they "bind." The arms races between parasites and their hosts are all about these binding proteins. Parasites invent new keys; hosts change the locks. There is an obvious group-selectionist argument here for sex: at any one time a sexual species will have lots of different locks; members of an asexual one will all have the same locks. So a parasite with the right key will quickly exterminate the asexual species but not the sexual one.*[110]

Reproduction by fission or cloning produces offspring identical to the parents. With cloning the "locks" stay the same from generation to generation. If the parents are vulnerable to parasites, or in the process of becoming vulnerable to parasites, cloned offspring are just as vulnerable.

The greater a parasite's familiarity with its host's genes, the better it is at squeezing out host resources.[111] But by reproducing sexually, an organism can switch out many of its genes. This produces young that are genetically different from their parents. Critically, it does this without completely disrupting hereditary transmission of the parents' DNA line. Sex puts up a roadblock against parasites' entry into the next generation. Sexually reproducing parents' "somewhat different" offspring will be unfamiliar to their parasites, and therefore less likely to be efficiently exploited.

Pool game

This "evasion through sexual recombination" trick depends upon genetic diversity. A group of eukaryotic animals free to mate with each other is called a breeding population. Every individual in that group contains two copies of every gene. There are many variants of each gene. The variants are called "alleles" of the same gene. One individual may have two of the same allele, or it may have two different alleles. A breeding population is a gigantic storehouse of alternative alleles, a "gene pool." The capacity of this storehouse is many, many times the storage capacity of a single individual. By combining sexually the individuals in a breeding population take advantage of this rich collection of alternative genes.

Sex keeps the host's parasitic opponents off balance by drawing from the gene pool to create new combinations of genes. These fresh mixes befuddle the parasite, which can never fully adapt. Returning to the metaphor of button-pressing, what is a "right" set of buttons to press now will be a "wrong" set in the next generation. This shell game makes the host slippery enough that the parasite can't maintain its traction.

Neither archaea nor bacteria can pull this trick off. Their asexual reproductive strategy means they don't have this kind of gene pool. Although bacteria do have the ability to "borrow" genes from their bacterial neighbors, this is a hit-or-miss process. There's no way for them to maintain the genetic diversity that would be required to rapidly remix their genes. Sexual species, on the other hand, have at their disposal an orderly "lending library" of alternative genes. As Ridley explains,

> *Sexual species can call on a sort of library of locks that is unavailable to asexual species. This library is known by two long*

words that mean roughly the same thing: heterozygosity and polymorphism. They are the things that animals lose when their lineage becomes inbred. What they mean is that in the population at large (polymorphism) and in each individual as well (heterozygosity) there are different versions of the same gene at any one time.[112]

Nick Lane further clarifies the logic of this defensive tactic.

Why be different from your parents? .... If the host population is genetically identical, then the successful parasite has the run of the entire population and it may well be obliterated. If the hosts vary among themselves, however, there is a chance, indeed a probability, that some individuals will have a rare genotype that happens to resist the parasite. They will thrive until the parasite is obliged to focus its attention on this new genotype or face extinction itself.... So sex exists to keep parasites at bay.[113]

There's evidence to support this interpretation. Certain eukaryotes – for example some species of snail – are able to reproduce both sexually and asexually. The conditions around them determine which method they will employ at any given time. If it is true that sex gives resistance to parasitic exploitation, you'd expect these snails to opt for sex more often in environments where parasites are more active. So researchers have compared rates of snail sexual reproduction in environments of heavy parasite infestation versus environments of less heavy infestation.

Researchers King et al studied populations of such snails (a species named *Potamopyrgus antipodarum* to be specific) in their natural stream habitats. They found that living in parasite-infested waters did indeed tip the scales toward sexual reproduction. Further, they noticed that even when the snails reproduced by cloning, their genetic diversity was greater where parasites were more active. This also supports the theory. Higher diversity among clones would make sense if genetic diversity really does provide valuable protection against parasites.

Under the Red Queen hypothesis, host-parasite coevolution selects against common host genotypes... Clonal diversity and the frequency of sexual individuals were both positively related to infection frequency. Surprisingly, although clones are derived by mutation from sexual snails, parasites explained more of the genotypic variation among parthenogenetic subpopulations. Our findings thus highlight the importance of parasites as drivers of clonal diversity, as well as sex.[114]

So it does seem that for those snails that have an option, living in the presence of parasites leads them to choose sexual reproduction over simple splitting. And even where they don't use sex, living with parasites increases the diversity in their genes. This kind of data adds credibility to the Red Queen hypothesis.

As we noted earlier, not every scientist agrees that sex evolved as a defense against parasites. But at least for now this explanation seems to explain available data better than any alternative.

Simply irresistible

So sex is an extreme defensive measure put together by evolutionary forces. It came into being as a way of addressing the very difficult problem of parasitic infestation. But sex introduced problems of its own.

A significant complication of sex is the need for a mate. When organisms reproduce by splitting, they already have everything they need to make copies of themselves. They can create offspring simply by copying their own DNA. But when an animal is going to mate, it has to locate a partner of the opposite sex. That individual is generally not near at hand. With any luck a partner is out there somewhere. But how do male and female get together? Sexual animals have adopted a bewildering variety of methods for hooking up with that "special someone."

Animals of many species use sex pheromones, chemical scents that broadcast location and readiness to mate.[115] [116] Lots of plants and aquatic animals use a shotgun approach, peppering the local environment with so many DNA packages that some are likely to find the opposite sex.[117] [118] Flowering plants have found a way to bring insects and birds into the loop as carriers of their pollen.[119] Insects such as cicadas and crickets use characteristic sounds to guide their mates to them.[120] Some insects use chemically-produced flashing lights.[121] Many birds use songs.[122] [123] Whatever the specific form, mating activities are complicated and take a lot of time and energy.

Humans have their own convoluted methods, the details of which vary from culture to culture. Upon reaching reproductive age males and females begin to experience longings for the company of the opposite sex, and to send out "sexual signals" about their openness to mating. These consist of changes in bodily appearance reflecting sexual maturity, health and fertility, possession of resources for childrearing, signs of acceptable cultural conditioning such as current fashion, and behavior meeting standards of desirability and "coolness" within the subgroup with which the individual is identified.

Does the mating game end when a potential partner is located? Not hardly. In many animal species (including our own) a meeting triggers a complex ritual of critical evaluation. To illustrate, in many species of bird the male goes through very close visual inspection by the female. In addition the male must precisely perform a set of behavioral gyrations, a complex "dance" that must be executed flawlessly before the female will allow any further progress toward mating. It seems the purpose of the inspection and rituals is to ensure a choice of the strongest, most capable, healthiest and most parasite-free male available.

Like other animals, we humans have strong feelings and drives that guide us in sexual matters. Choices are generally not made by thinking things through. Instead, they are guided mostly by automatic emotional responses. These emotions and drives are themselves products of evolution. Attraction is, as they say, a matter of the heart rather than of the head. When a man spots an appropriate mate, he may unwillingly experience that person as "stunning," "gorgeous," "dead sexy,"[124] "hot," or "simply irresistible." A woman may experience a man of her choice as "sexy," "good looking," "hot" (or, confusingly, "cool"), or "yummy." Our potential mates may look, move, speak, smell, and behave in ways that implode our defenses. They bring out the animal in us. They take our breath away. We find that something about them, something we may not be able to fathom, just drives us wild. We don't always understand our choices, and may not be able to justify them rationally. We may be distressed to find ourselves bonded sexually and emotionally to someone we don't really like (or worse, someone who doesn't like us).

Conversely, our rejection of a potential mate comes to us as an involuntary negative emotion – for example, we may experience those we reject as "gross" or "repulsive." Or they may simply leave us cold.

Sexual attraction and attachment often defy our ordinary understanding. But a close examination shows these things don't happen without a reason. Science says sexual attraction has deep roots in evolutionary biology. The features that men and women find attractive in each other relate directly to the likelihood of successful reproduction. Though the drives and emotions we experience may not be "logical" in our usual sense of that word, they do make sense in the context of what has made for successful reproduction through evolutionary history.

The evolution of enchantment

How can we best talk about sexual and romantic attraction,

those mysterious forces that add such depth of meaning to our lives? What is it to be smitten by love? Why are lust and romantic feelings so deep, so meaningful, and so likely to unbalance us? Can science give us answers?

Science does have some answers. But to get a sense of our subject matter we might do well to start with literature, because it has long tread sacred ground that science has been reluctant to trespass. And what better literature than the classics. Romeo has this to say about his beloved Juliet in the second act of Shakespeare's play *Romeo and Juliet*:

> *But soft, what light through yonder window breaks?*
> *It is the east, and Juliet is the sun.*
> *Arise, fair sun, and kill the envious moon,*
> *Who is already sick and pale with grief*
> *That thou, her maid, art far more fair than she.*
> *Her vestal livery is but sick and green,*
> *And none but fools do wear it. Cast it off!*
> *It is my lady. Oh, it is my love.*
> *Oh, that she knew she were!*[125]

Students often interpret this passage as a flowery tribute to Juliet's beauty. Well, yes. But it's more than that. The language is archaic, and the meaning of Shakespeare's subtle metaphors often eludes modern audiences. So here's a breakdown of the action.

From a hiding place Romeo is watching Juliet as she stands on her balcony. Although it is dark outside, Romeo's eyes have been opened to an ethereal brilliance. Emanating from that balcony is the blinding light of Juliet's beauty. Romeo is smitten by love, you see. Enchanted by her surpassing good looks, he likens Juliet to the sun driving away the darkness. But with frustration he imagines her as a handmaiden to the virginal moon goddess Diana. The moon goddess is "sick and pale with grief" in envy of the radiance of his Juliet. Romeo feels it is high time for Juliet to break rank with the virgins, and become his lover. He argues in his heart that (since Diana's kind of pissed off by Juliet's beauty anyway) she should go ahead and dump the goddess of virginity. Especially because lengthy maidenhood is rumored to be both foolish and bad for the health. Juliet is his chosen love, he vows. And he wishes she knew just how much he cares about (and lusts after) her.[126]

James Joyce, in his sensational 1918 work *Ulysses,* did a mesmerizing job of portraying simultaneously the verbal and the more deeply emotional aspects of his character's sexual desire.

Yes when I put the rose in my hair like the Andalusian girls used or shall I wear a red yes and how he kissed me under the Moorish wall and I thought well as well him as another and then I asked him with my eyes to ask again yes and then he asked me would I yes to say yes my mountain flower and first I put my arms around him yes and drew him down to me so he could feel my breasts all perfume yes and his heart was going like mad and yes I said yes I will yes.[127]

Descriptions of sexual fixation are certainly not limited to the classics. Today a wealth of romance literature captures the flavor of human sexual attraction, perhaps more authentically than does science. Here's an excerpt from a modern romance novel:

Unable to speak, his eyes did all the talking for him, sweeping over the soft features of her face, her graceful neck and the slender curves accentuated by her light summer dress, short enough to show off her shapely legs all the way down to her elegant ankles. He noticed her chest rise with every breath she took... How long he'd stared at her, Daniel truly couldn't tell. Maybe a second, or maybe as long as five minutes. But he knew why he was suddenly tongue-tied. It was a clear case of lust. Severe lust. Uncontrollable lust.[128]

Yes, yes. Romance. Love. Lust. The intense, involuntary, persistent emotional attachment that leads to sexual union. Where people are concerned that attachment is triggered in a highly complex way that includes visual, physical, behavioral, and cultural cues. Science is finally sniffing among those cues. But it is as yet hampered by caution about delving into such an emotionally charged area of human life.

What has science discovered so far? That (predictably) romantic and sexual attraction have their roots in evolutionary pressures, including (perhaps surprisingly) the pressures created by parasites.

Sexy science

Darwin was impressed by the great diversity across cultures in the things people find sexually attractive. He concluded that humans don't really have biologically-driven beauty standards. But since then science has shown he was mistaken. There are definite links.

Summarizing research, Henderson and Anglin point to six things that humans from all cultures find universally attractive. These things are youth, facial averages (how closely the individual resembles the average face within a culture), body symmetry, prominent secondary sexual features (visible evidence of gender), body type, and perceived health. Individuals that measure up on these dimensions are heirs to "a cornucopia of benefits."[129] [130] These benefits include greater likelihood of attracting mates, and greater health and longevity.

During the last 15 years or so, multiple studies have documented the benefits of facial and body attractiveness. A study by Shackleford and Larsen found that facial attractiveness actually does reflect health.[131] Henderson and Anglin conducted a study to see whether facial attractiveness related to how long a person would live. Ratings of a person's attractiveness as an adolescent did predict how long that person would live, especially when the raters were male.[132]

It may be surprising that attractiveness would be linked to

health and longevity. Why would there be such a connection? Well, it does make sense when you think about it.

When you hook up with a healthy mate your offspring are more likely to be healthy themselves. Because they are healthy, they are more likely to survive. Many generations ago, some individuals were (perhaps by chance) more attracted to facial regularity, symmetry of features, and other indicators of health. These lucky individuals left more descendants. And because they left descendants their chance attraction was passed on. Over the generations the experience of attraction was further amplified and refined. Now people with these features just look "hot" to us.

Conversely, some other individuals weren't attracted to signs of health. They ended up with unhealthier mates. They had unhealthier children. These weaker children were more likely to die without leaving heirs. This created a selection pressure. And what was being selected for was the experience of being turned on by signs of health.

And so in the fullness of time evolution translated the ability to read signs of health into the subjective experience of attraction toward these features. We are heirs to this evolutionary legacy. We directly experience these signs as a turn-on. We find ourselves involuntarily captivated by youth, liveliness, facial symmetry, clear skin, and so forth. These things are just sexy. We can't help it. No matter what our intention, we get aroused by and drawn to the very things that enabled the continuation of our ancestors' line.

Now, in the present, individuals with such "attractive" features enjoy an advantage in the mating game.[133] Researchers Rhodes et al wanted to find out whether facial attractiveness of men and women influenced the level of their sexual activity. Their findings match our commonsense expectations.

Men with attractive faces and bodies enjoyed significantly more short-term mating success than their peers, with no cost in their long-term mating success, whereas women with attractive faces had more long-term mating success than their peers. Attractive men (bodies) and women (faces) also became sexually active earlier than their peers, which would enhance reproductive success for both sexes in the absence of contraception. These results support the assumption underlying much current research on attractiveness, that attractive traits are important and mate choice and may be sexually selected.[134]

These researchers noted that to women, "masculine" faces "honestly advertise health during adolescence." Men's facial masculinity and symmetry both improve their mating success. Women's facial beauty seems more important in attracting male sexual partners than does beauty of the body.[135]

According to researchers Singh and Young, men of most societies are turned on by plumper rather than slimmer women. Where food is scarce and illness is common, plumpness shouts health. But the message given by that body fat depends on its *location* on the woman's body. Wide hips make for easier childbirth. For health fat should be on hips and breasts rather than on the waist. Waist-to-hip ratio is a better predictor for freedom from disease and good sex hormone balance than either overall body fat or breast and buttock size. So waist-to-hip ratio counts more in female attractiveness than overall body weight.[136] Studies by these researchers confirm that even in the U.S., where slender women are considered most attractive, the right waist-to-hip ratio presses men's sexual buttons. They conclude that

[L]arge-breasted figures with low WHR [waist-to-hip ratio] are judged as highly attractive, feminine looking, and healthy, and are preferred for both short- and long-term relationships provided that such figures are slender and do not have large hips.[137]

They emphasize that the most likely reason for this turn-on is that the waist-to-hip ratio is strongly related to a woman's health.

[W]aist size is significantly correlated with many major diseases such as cardiovascular disorders, diabetes, and gallbladder problems...The high positive correlation between attractiveness and healthiness supports the belief that health may be a defining feature of attractiveness. The health status of a woman would be important to men whether seeking a casual or a serious relationship.[138]

Parasites and peacocks

In the end, everything about our sexual attraction boils down to having children and seeing them raised to maturity. Mind you, our judgments of attractiveness are not made consciously, by thinking it through. Rather, they are made through involuntary emotional experiences of romantic and sexual longing.

Researchers Townsend and Wasserman did a series of experiments to sort out the things that were attractive to college men and women. Because all the subjects were in college, their findings may not fully apply to non-college men and women. Even so, they paint a familiar picture. These researchers found that both men and women "screen" potential partners to decide whether they would consider dating or having sex with them. The screening takes place at both conscious and unconscious levels. Only when a potential partner passes this first screening do other factors come into play.[139]

To decide whether they would consider dating a man, women screen for nonphysical characteristics. They first pay attention to things that reflect a man's potential for investment in her future children. These include ambition, income, social status, and rank within the woman's local social group. If a man doesn't measure up in those areas, she is more likely to rule him out as a potential partner.

For women, a positive initial assessment, which is also influenced by partners' physical attributes, typically opens the door to a first date. A first date for women is a chance to explore partners' potential for higher-investment relationships, e.g. partners' warmth, values, family background, SES, and interest in them.[140]

If a man meets her basic expectations, the woman considers other things – like bodily and facial features, personality, and evidence of the man's actual or possible interest in her. "Given the potential costs and benefits of short-term mating for women... women tend to test and evaluate short-term sex partners as long-term prospects rather than being end goals themselves. Consequently, women's criteria for short-and long-term partners are similar..."[141] Physical characteristics are important for women. But good looks influence their choices most within the pool of men she has already pre-selected as having potential. In other words, appearance starts to count when the woman is considering a guy she's already judged has something to offer.[142]

Men seem to make their choices very differently. They use good looks for their first screening of women. "For men, potential partners' physical attributes largely determine the pool of partners with whom they desire sexual relations," say these researchers. "Partners' physical attributes thus establish a pool of partners who are acceptable for sexual relations and who may merit further investment."[143]

It is true that – for both men and women – high status can make up for lack of good looks, and good looks can partly make up for low status. But these trade-offs work differently for the different sexes.

For middle-class men, high physical attractiveness can render women desirable for dating, sexual relationships, and even marriage regardless of their occupation, income, and education – provided that they do not exhibit the obvious trappings of a lower-class status and lifestyle. In comparison, women appear unwilling to date, marry, or have sexual relations with low-income, uneducated males regardless of the men's physiognomies and physiques.[144]

Good health is essential for reproductive success. So indications of good health are a turn on for both males and females. One researcher has used a mathematical model to show that if a species ignored indications of health in choosing mates, that species would promptly go out of existence.[145] Males of many species use some form of flamboyant ornamentation in their courtship rituals. It has become clear that these displays directly advertise health and fitness. The most dramatic example is the peacock, whose gigantic and exuberantly multicolored tail clearly taxes his ability to manage it.[146]

This is a book about the influence of parasites. So at this point it will perhaps come as no surprise that males of many species use sexual displays to advertise their parasite status to females. The notion that a male's ornamentation advertises its ability to resist parasites was first put forth by researchers Hamilton and Zuk.[147] According to researchers Saks et al, those males that are in good health generally are more resistant to parasites, and so are better able to indulge in extravagant displays to impress the ladies.[148] By taking advantage of these cues females are able to select the most parasite-resistant males available.[149]

This "parasite advertising" thing may seem a little strange. Are we sure it really happens? Yes, the scientific evidence is pretty strong at this point. For example, Saks et al studied the connection between health and coloration in green finches (*Carduelis chloris*). Their results showed that "carotenoid-based plumage coloration in green finches honestly signals immunocompetence and health status."[150] The same is true of a wide variety of other species, from field crickets (*Gryllus bimaculatus*, who advertise their parasite status in their songs)[151] to house mice (*Mus musculus domesticus*, who use the quality of their urine as a token for parasite status)[152] to blackbirds (*Turdus merula*, who use the intensity of color of their bills)[153] to Asian elephants (*Elephas maximus*, who signal parasite status with the size of their tusks).[154] Males' broadcasting of their parasite status is so widespread among species that it just has to raise the question of whether human males do something similar. It turns out that they do.

Testing for testosterone

Human males do advertise their parasite status to females. One of men's involuntary signals involves the visible characteristics we call "masculinity." High levels of the male hormone testosterone have physical effects on men's bodies, giving them more "manly" features. Testosterone induces multiple bodily and behavioral changes. Bodybuilders artificially raise their testosterone levels to encourage the development of muscular physiques. Testosterone also changes their faces. For example their jaws get larger and more prominent, as do the ridges under their eyebrows. And as we know, higher testosterone levels are associated with aggressiveness and dominance in males.

Testosterone has its downside. Researchers DeBruine et al emphasize that high levels of testosterone are known to weaken immunity. To support higher testosterone levels men have to have better-than-average immunity to begin with. In this sense it's like the peacock's magnificent but burdensome tail. To support the extra testosterone they need to have been blessed with greater resistance to disease. The bottom line is that "masculine" appearance becomes an honest indicator of parasitic resistance. Women can choose males with good resistance simply by choosing those with manly appearance.[155]

There's a catch. High testosterone makes men "testy" in temperament. They're more difficult to live with, are poorer bets for quality family life, and are more likely to stray sexually. "Indeed," say these researchers, "there is compelling evidence that women ascribe anti-social traits and behaviours to masculine men. Women perceive masculine men as dishonest, uncooperative, more interested in short-term than long-term relationships, and even as 'bad parents.'"[156]

Researchers Gangestad and Scheyd noted that women are more turned on by "manly" appearance and behaviors at those times of the month when they are most likely to conceive. This short attraction applies only to consideration of a man as a sexual partner — not to his attractiveness as a long-term mate.[157] The monthly "man-candy" excitation might work to give a woman healthy children without the handicap of a long-term relationship with that troublesome, high-testosterone male.

Given the negatives of the high-test male, women have reason *not* to prefer them as mates. So there must be some kind of biologically-based incentive for fooling around with manly men.

DeBruine et al reasoned that health could provide that incentive. Health, they thought, would be a bigger issue for women in countries with poorer average health. They predicted that the worse a country's health index, the more important the testosterone-related signs of a man's good parasite resistance would be. Being more concerned with health, the women living in health-poor countries[158] would prefer men with visible signs of high testosterone. That is, they would be more attracted to "masculine-looking" men.

This hypothesis is testable. DeBruine et al collected sexual preference data from over 4,500 women across multiple countries and cultures. They showed them pictures of men's faces that were more feminized or masculinized, and asked the women to rate their attractiveness. The results?

> *We found that cross-cultural variation in women's average masculinity preference was predicted by a NHI [national health index] derived from eight World Health Organization statistics for mortality rates, life expectancies and the impact of communicable disease. Consistent with predictions from sexual*

selection theory, as national health decreased, women's average masculinity preference increased...Across cultures, as national health decreases, women's preferences for masculine men increase.[159]

A second set of researchers (Lee and Zietsch) studied the disease/attraction connection using an entirely different approach. They had a group of American women subjects fill out a questionnaire about diseases. This task brought the issue of disease to the forefront of the women's minds. Then they measured the women's preferences for the faces of artificially feminized or masculinized men. Women who had been sensitized to disease tended to prefer masculine faces more than women who had not been sensitized this way.[160]

Other researchers (Gangestad and Buss) were interested not just in manliness, but in the physical beauty of both sexes. They performed a large cross-cultural study of attraction. They found that both men's and women's preference for physical beauty went up as the level of parasites in the environment rose. They write "Human data from 29 cultures indicate that people in geographical areas carrying relatively greater prevalences of pathogens value a mate's physical attractiveness more than people in areas with relatively little pathogen incidence."[161]

Not only attraction, but also revulsion has been tied to concern with disease resistance in a potential mate. Researchers Park et al note that one function of humans' experience of disgust is to protect them from potential carriers of disease.[162] They conducted a series of studies to see whether disgust with pathogens affected perceptions of attractiveness. They found that unattractive faces are judged especially repellent by men and women who are strongly repelled by disease. These researchers suggest that the experience of revulsion may be a tool created by evolution for protection in environments where pathogens are common.[163]

All these studies support the idea that physical appearance plays a bigger part in sexual attraction where disease is prevalent. Though we may not be aware of it, the attraction (or revulsion) we feel for our potential mates is rooted at least partly in the need to fight parasites.

In Summary

Sex isn't the only way to make babies. Sexual reproduction didn't even exist until it was invented by eukaryotes. The available data says sex isn't the best way of passing genes to new generations – and speaking objectively it doesn't even seem to be a very good way. In fact it's so inefficient that in a "clean" environment it falls out of favor. So why do almost all eukaryotes reproduce this way?

Apparently sex is a super way to fight parasites. It really shines where the environment is full of them. The fact that almost all eukaryotic organisms reproduce sexually is testimony to the fact that their typical environment has lots of parasites. Sexual recombination shuffles our gene alleles, confounding disease organisms. Further, sexual attraction is finely attuned to signs of ability to resist parasites. We're involuntarily turned on by those physical features that provide cues for high resistance to parasites. Nature's driving the bus here – we're passengers on a scenic tour. The greater the prevalence of parasites around us, the greater the chances we'll be blown away, slapped upside the head by the physical beauty (and other health indicators) of our potential mates.

From this perspective we can see that sex belongs in the same basket of tools as innate and adaptive immunity. Both immunity and sex are weapons used in the fight against parasites. The difference is that the focus of immunity is on protecting the individual organism, where sex serves to protect the entire gene pool.[164]

Chapter Four
Death and parasites

C hapter 3 outlined a plausible explanation for sexual reproduction. The best evidence we have says a major purpose of sexual reproduction is to shuffle gene alleles. Animals use gene shuffling to keep their parasites confused, and so prevent them from getting out of control.

Reaper wanted, apply within

But if we accept this logic we are confronted with a problem. With each new generation, sexual reproduction whips up "new model" animals having fresh and unique gene combinations. What about the "old model" animals – members of the parent generation? Is it okay if they stick around to intermingle with the newbies? Could that be a problem?

Well, the gene-shuffling rationale for sex suggests that keeping the old-timers around would weaken the system. That's because parasites will have adjusted themselves to the older generation. The continued existence of the older, parasite-friendly DNA patterns would be slow poison for the Red Queen. The longer the parents live, the more progress their parasites will make toward winning the continuing arms race.

If shuffling is truly the explanation for sex, it follows that the "old models" must somehow be given the boot, taken off the game board. The rationale for sexual reproduction implies a role for a Grim Reaper – a way to reliably erase the older DNA packages from the picture.

Level up!

One of the most disturbing things about our personal lives is the certainty of our deaths. The inevitability of our passing is so familiar to us that we rarely question it, and may prefer not to dwell upon it. But do we really understand death? When we take a closer look, we will find that senescence and death are all about protecting the breeding population as a whole.

With the coming of cellular differentiation and true multicellularity, the unit of natural selection changed. It went to a new level. Individual cells were no longer the unit of selection. Now that unit was the multicelled organism as a whole.

With entire multicelled organisms being selected, is it a big deal when individual cells within the body die? No. Death of an individual cell no longer means death of that cell's DNA package. Even when a cell is dead, its DNA can still be passed on by the germ cells of the animal. In terms of DNA the cells in the body are clones of each other. So any given cell within the body can be stone cold dead without, so to speak, "passing away."

The animal's germ cells serve as proxy for each of its individual cells, dead or alive. If the death of an individual cell serves a useful purpose for the organism as a whole, helping the larger animal to survive, then that cell is a "fit" cell. It is fit even if deceased. At this level, "fitness" is a word that only has meaning in relation to the health of the organism as a whole. The death of an individual cell can support the life of the larger animal. Multicelled animals are selected according to their fitness as a team of differentiated cells. The only important issue is whether or not the DNA of the team is passed on to the next generation.

But wait! Sexual recombination of genes prevents the multicelled "fitness winners" from passing their complete DNA package to the next generation. Instead, winning individuals simply contribute parts of their DNA package to the gene pool shared by the breeding population. What good is it to be a "fitness winner" when you can't directly pass on your DNA?

With the coming of sexual reproduction the focus of evolution moved even further away from the individual cell. In fact that focus moved largely to the gene pool, with its extensive store of genes. The gene pool, housing the "crown jewels" of the species, became the most important thing passed on from generation to generation.

With this change in the level of evolutionary focus, the survival of any individual animal isn't a show-stopping issue either. Death of any individual multicelled animal does not mean oblivion for that individual's genetic information. If an animal dies without reproducing, the individual gene alleles that made it up will likely continue to exist within the gene pool (though at reduced frequency).

Again, when an animal survives to reproduce sexually it doesn't get to replicate its complete gene package. The "most fit" sexual animals are never allowed to reproduce themselves completely or exactly, as is the case with prokaryotes. That restriction is what makes sex a weapon against parasites. With sex, the influence of an individual's gene package is strongest in his or her own immediate offspring. But this direct influence is progressively diluted by mating within each succeeding generation. After a time, the sexual contribution of a single "fitness winner" has served only to ratchet slightly upward the frequency of its gene alleles within the breeding population as a whole. So in this sense sex ensures that what is ultimately being selected is the gene pool. This way of thinking is called Population Genetics.[165] [166]

The immortal bacterium

As we have seen, teasing out the adaptive benefits of sex was a pretty confusing project. Getting clear about how death can improve survival twists the brain even more cruelly. Nevertheless programmed death (death initiated by the organism's own genes) can be understood as a sensible strategy for managing parasitic infection. So programmed death also belongs under the larger umbrella of immunity.

Before sex was available, organisms could reproduce only by splitting ("cell fission"). Before it splits, a cell's DNA is copied. Then during the physical split of the cell one half of its resources, including one of the two DNA copies, is allotted to each of the two daughter cells. Afterwards each of the daughter cells is the spitting image of the parent – genetically and bodily. This way of reproducing started a long time ago. Assuming adequate environmental resources, it will continue far into the future.

It is essential that we distinguish between programmed death (that is, termination of life under genetic control) from the simple cessation of life. Let's signify programmed death by capitalizing it like this: Death. The life functions of individual bacterial cells do come to an incidental end quite often – this happens all the time. But it's not because they "get old and die." They generally cease to live only when they get hurt or run out of resources – when they have reached an environmental dead end. They find themselves in a place where there aren't enough resources to sustain their lives. But coming to an end this way isn't the same as Death. Bacteria don't slow down and get senile and crotchety. They don't develop aches and pains, and then keel over.

All bacterial lines existing today are very, very old. And there's no sign that they're wearing out. Unlike us, these cell lines are in a meaningful way immortal. There is a sense in which new generations do not replace their ancestors – rather, they *are* their ancestors. Every prokaryotic line alive today has been splitting and splitting and splitting for eons.[167]

These bacteria entered into their immortality billions of years ago, and they are still around. To be sure they have been modified, and continue to be modified, by evolutionary pressures. But they are nonetheless still kicking. Literally older than the hills, such cell lines seem the picture of bacterial health and vitality. Despite their being ancient even in a geological sense, they aren't "getting old."

Isn't that impossible? How do they do it? Don't bacteria experience chemical damage over time like our own cells do? Well, they do. But bacteria have the ability to "slough off" damage as they split.

Authors Aufstad et al describe a recent study of the bacterium *Escherichia coli*. These bacteria are sausage shaped. In splitting, a cell lengthens until it is about twice its original length. Then it pinches itself off in the middle to form two daughter cells. The researchers found that when cells divided, damaged proteins within the original cell are herded into the oldest of the two segments. The result is one "brand new," rejuvenated cell and one less functional cell having some damage. The less fit cell may be less likely to survive. But the other is "brand new," good to go for another generation. The authors believe that this "rejuvenating reproduction" is likely the norm among bacteria.[168]

Back when life began, immortality was all there was. It was the norm because there weren't any alternatives. Even today immortality remains the normal state for prokaryotes and archaea. But my life, your life… that's a different story. We high-rolling eukaryotic organisms do in some ways seem more sophisticated. But our lives will predictably end in dwindling energy, decay of bodily function, creeping senescence, and rattling, essence-imploding Death. It's natural to wonder why this has to be. Or even if it has to be.

The continuing vigor of ancient bacterial forms is perplexing. After all, if all living things must wear out, you'd think after a couple of billion years bacteria would be *really* worn out. So their continued vitality hints there's something about aging and Death that we don't understand. Let's take a closer look.

Old before their time

The conventional wisdom is that "wearing out" causes senescence and Death. It is said that like a car, our body is great fun to drive around while it's new. But after awhile it starts needing repairs. Over the years it goes into the shop more and more often. Before you know it, your beloved vehicle is a junker barely able to keep up with traffic. And one sad day it moves on to that great salvage yard in the sky. Death. Auto parts don't last forever, you know. And neither do our bodies – or so we're told.

But there's something about this "wearing out" theory that doesn't quite ring true. Consider the disorder called Progeria. Children with this illness seem to grow old way, way before they've had time to wear out. Hutchinson-Gilford Progeria Syndrome is caused by a rare, fatal genetic abnormality. That abnormality causes excess secretion of a protein called "progerin" (from the Greek "pro-" meaning before or for, and "geras" meaning old age). Beginning almost immediately after birth, the child begins to "get old" at an astonishing rate – about 10 times normal speed.

For example, the child's face takes on a characteristic "elderly" appearance with small, recessed jaw, beak-like nose, and bags under protruding eyes. His hair falls out. His skin gets thin, wrinkled, and mottled. His bones thin, and his body fat wastes away. He develops a small fragile frame with stiff, easily displaced joints. He is prone to developing atherosclerosis, and will likely die from heart problems or other complications of atherosclerosis. In short, almost every feature we associate with the elderly is exhibited by a child only a few years old. The average age of a progeric child at death is about 12 years.[169]

Recent study of Progeria has led to some startling discoveries. First, progerin appears to be a normal component of *all* human bodies – not just the bodies of these sick children. There's just more of it in the bodies of progeric kids. Second, progerin levels seem to play a key role a well-coordinated, normal sequence of senescence and death in the human body. Progerin affects multiple genes and enzyme systems in the body, in essence converting a regular person into an old person. Third, within individual cells the expression of progerin seems to be triggered by shortening of that cell's telomeres past a critical point.[170]

Telomeres are the "end caps" on each of our chromosomes. Every time a cell divides, this end cap is shortened. If the cell's telomeres get too short it can't divide anymore. At this point the cell starts expressing progerin in abundance. It's shouting to the whole body "I'm old!" When enough cells are yelling the same message at the same time, it seems, the senescence genes are turned on. We are transformed bodily into old folks.

Francis Collins is the director of the National Institutes for Health. He and his team have studied Progeria closely. At a recent talk he emphasized that as we age progerin is being made continuously in a fraction of our cells – but at much lower levels than in progeric children. "Collins' lab team found using cells grown in the lab, when they used a chemical to uncap the chromosome ends, it altered gene splicing and led to progerin production and cellular senescence." Once this process is started, they found, those cells produce progerin at the same high levels as in the progeric kids. Collins commented, "While I can't prove it, it seems likely that progerin itself is part of normal, programmed senescence."[171]

Recently a large-scale human study was conducted to further

assess the reported link between telomere length and health. The average age of the 100,000 participants was 63 years. Researchers described the study at a recent conference.

> *"We found that individuals whose telomeres were in the shortest 10 percent were about 23 percent more likely to die in the three years following measurement of their telomeres, when compared with individuals whose telomeres were longer," said lead study author Catherine Schaefer... This was true even after adjusting for the effects of smoking, alcohol consumption, education and other factors that are associated with telomere length, the study found. The findings may suggest that shorter telomere length is not just a byproduct of the aging process but, instead, perhaps itself a significant root cause of aging and death, the researchers said.*[172]

This line of thinking is supported by an additional observation about so-called "immortalized" human cancer cells. In culture some cancerous cells keep reproducing themselves without any apparent limit. That's because they are manufacturing an enzyme called telomerase. This enzyme keeps rebuilding the cell's telomeres, so they don't get any shorter. In this way these cancer cells do an end run around the normal trigger for senescence.

If there's any biological law that says "cells must get old and die" these cancer cells are completely ignoring it. Interestingly, these rebellious cells never express progerin. One research team found that "In support of the notion that progerin production is selectively induced in senescent cells with short dysfunctional telomeres, we showed that progerin mRNA [a mid step in progerin production] was essentially absent in immortalized cells."[173]

What's this all about? As we have seen, the purpose of sex may be to shuffle the genes so as to provide resistance to parasites. If this is indeed the explanation for sex, then we can understand the value of Death following sexual reproduction. Sexual reproduction creates a new generation of parasite-resistant shuffled-gene models. Now the old vulnerable models are obsolete. They have to go. Their sticking around would only give the parasites a comfortable home, making it easier to survive while those dastardly creatures diabolically plan (OK mindlessly evolve) their next adaptation.

D-E-A-T-H spells life

How did eukaryotes acquire Death? The evidence suggests they inherited it from bacteria, who stole it from viruses. Programmed death was apparently first used by certain types of sea-dwelling prokaryotes. This happened a long time ago – billions of years. It seems these bacteria stole Death from the viruses that were feasting on them, and turned their weapon against them. They started using self-induced Death as a way of managing those viruses.

Ocean-dwelling prokaryotes periodically have gigantic "blooms" that cover hundreds of square miles of water. They did the same thing billions of years ago. Each of the individual organisms in these blooms is genetically the same, because they reproduce by splitting. The organisms in the bloom are the most successful of the current generation of bacteria – the "winners."

As the number of bloom organisms soars, viruses called phages are increasingly able take advantage of their large numbers and their ultra sameness. Winter et al note that "viruses are abundant and active members of aquatic food webs and constitute a major source of mortality for prokaryotes."[174] The burgeoning numbers in the bloom allow these phages to acquaint themselves with the strain very well, gene by gene. This intimacy allows them to completely take over their hosts.

While the bloom is in progress the phages are busy reproducing themselves inside the host bacteria. When a large number of new virus particles have been assembled, the virus makes the host cells disintegrate. This is called lysis. The virus makes lysis happen with special enzymes called caspases. The cell disintegration releases the newly manufactured viruses into the seawater. The phages kill off pretty much the entire population in the bloom at the same time, releasing unimaginable numbers of virus. This has been termed the "kill-the-winner" viral strategy.

It isn't hard to understand that killing off the bloom to release new viruses benefits the phages. Oddly, though, kill-the-winner seems to benefit the host organisms as well. It has the effect of ensuring that multiple strains of the same host bacterium stay alive. It allows bacteria, organisms that don't have space for many genes in their own DNA string, to maintain genetic diversity within the larger bacterial population. This makes for a sort of "poor man's gene pool." The diversity partially makes up for not having gene alleles like eukaryotes do.

According to researchers Angly et al, the end result is that "as one microbial strain becomes dominant, its viral predator kills it and leaves open a niche that can be used by a related strain that is resistant to the phage... This model may explain the enormous microdiversity observed in microbial communities."[175] Authors Lopez-Garcia and Moreira make the same point. "Indeed," they say, "the strong demographic decrease caused in dominant cell populations by viral lysis (the so-called kill-the-winner mechanism) permits other, less competitive species to coexist at intermediate frequencies, resulting in the persistence of a large variety of species..."[176]

The kill-the-winner strategy of the phages ensures that there are always bacterial variant strains around, strains that differ from the current generation's winners.

The removal of the dominant strain makes room for the "somewhat different" strains – the ones that have been reproducing quietly on the sidelines – to move to center stage, thus forming the backbone of the next generation.[177] We can see that eukaryotes' strategy of taking old-timers out of the picture is not a recent development. It began with prokaryotes billions of years ago, before there were even any eukaryotes around.

Here's another benefit to the bacterial host: Winter et al say that when the bloom finally implodes, it releases into the ocean a mass of nutrients that serve to foster the development of the next generation of host "winners."

> *Viral lysis of host cells releases not only progeny virus particles but also a cocktail of sugars, proteins and peptides, amino acids, nucleic acids etc., that could serve as a source of nutrients for the surviving community.*[178]

Fighting over death

Angly et al emphasize a critical characteristic of the viruses that plague these prokaryotes: "Phages affect microbial evolution by inserting themselves into genomes."[179] The phages actually weave themselves into the DNA of their microbial hosts. Why is this important? It seems that once the killer caspase chemicals were introduced into the host DNA, the prokaryotic animals themselves started finding uses for them. This led to the invention of programmed cell death (PCD). A "chemical arms race" for control of the Death genes ensued.

Bidle and Vardi looked at the arms race between viruses and sea-dwelling algae. They note that phages' "tight control and co-opting of host PCD activation … are critical in order to prevent premature host PCD activation as a possible viral exclusion strategy."[180] Data they present imply that virus and host are locked in a life-or-death struggle.

But here's the twist: The virus and the algae are not struggling over *whether* the algae will disintegrate. They're both pushing for that. Rather, they are struggling over exactly *when* it will happen. They're fighting over the *timing* of the algae's suicide.

If host cells shut down early, the virus will not be able to replicate. The algae are motivated to kill themselves before the viruses have been fully formed inside them. But even as the algae are pulling the suicide trigger the viruses release special gene products into their hosts. These chemicals turn the unfortunate algae into "walking dead," zombies of a sort. They block the host algae from shutting vital functions down "long enough to ensure viral propagation."[181] Weinbauer et al make exactly the same point about cyanobacteria and their phages.[182]

A far, far better thing

A main character in Charles Dickens' novel *A Tale of Two Cities* is Sydney Carton. The novel concludes with the guillotining of Carton, who has altruistically arranged to give his life in order to save others. Carton's unspoken last thoughts include these famous lines: "It is a far, far better thing that I do, than I have ever done; it is a far, far better rest that I go to than I have ever known."

We have been examining organisms' use of voluntary Death to their advantage. Natural selection is the principle behind evolution. Does Death in any way help an individual organism, the presumed unit of natural selection? The idea that killing oneself could be "adaptive" may at first sound a little crazy. But with reflection it does make sense. Not sense for the individual, but for the population as a whole.

The strain of bacteria in a bloom is very closely related to those other strains that are not participating in that bloom. It turns out that when the bloomers kill themselves, more of their genes are passed on to future generations than would be the case if they didn't. How could this be?

We are used to judging an organism's fitness by its success in surviving to leave offspring. But there is the question of what eventually happens to those offspring. When does having lots of children fail to help you perpetuate your line? When the lives of all those children are sure to be snuffed out within a couple of generations! This is exactly the case with a bloom. One that kept right on blooming would be so over-run by viruses that it would be utterly wiped out. What's the difference between having trillions of dead descendants and having no descendants at all? From the evolutionary perspective, there isn't any difference!

Programmed death confers advantages to algae only because the various strains are closely related. When the genes of one strain are passed on, the vast majority of the genes of their relatives are passed on as well. Suppose *all* members of my immediate family are doomed to die. If I can save my more distant relatives (say my cousins) by killing myself, then it is in my genetic interest to do so.

Researchers Severin et al have documented the validity of this principle using populations of yeast cells, remarking that "the benefit of a population can outweigh the individual selective benefit if the individuals are highly related."[183]

By killing themselves at the appropriate time the sick ones give their more resistant cousins a chance to live on. If they didn't suicide, the whole species would quickly be overrun and wiped out by the parasitic phages. Like Sydney Carton, they give their lives that others dear to them might live. The suicide option may not be peaches and cream. But with viruses in the picture it may be the best option available.

This life-from-death principle applies even more strongly to the cells of a multicelled animal, whose DNA is not just highly related, but identical. Because the cells are clones of each other, the destruction of any specific cell doesn't hurt the multicellular lifeform as a whole. Whack a cell here, whack a cell there – it's no big deal. There are plenty more where those came from.

Reaper art

An old joke goes like this: A sculptor is famous for his fantastic portrayals of horses. An admirer of his work asks how he does it. The artist replies, "It's not that complicated. I start with a big block of marble, and chip away everything that doesn't look like a horse."

The body does something very similar. It readily sacrifices individual cells when their deaths promote survival of the animal as a whole. Here's an illustration: The most common form of programmed cell death is called apoptosis. Embryos couldn't develop into organisms without apoptosis. Why? Because the body "sculpts" its tissues and organs from shapeless blobs of cells. It does this by starting with way too many cells, and then killing off the ones it doesn't want.

Researchers Huh and Hay comment that "Cell death occurs throughout the development of every animal, enabling excess cells to be eliminated, tissues to be sculpted, and cells or tissues that have outlived their usefulness to be removed."[184] [185]

We've already looked at one example of the "sculpting with Death" principle in Lederberg's clonal deletion theory, the process by which the immune system learns the self-other distinction. Recall the way the body teaches the immune system to distinguish between self and other. It starts off with a lot more lymphocytes than it needs, then kills off all the ones that react to parts of the self. The ones that remain will be used to detect intruders. In fact the body uses "reaper art" extensively. As authors Entezari et al put it,

> *Embryonic development and differentiation to adult form depends on orchestration of cell division and death. In embryos, programmed death sculpts form, opens lumens [holes in tissue],*

separates or splits tissue layers, allows tissue layers to fuse and removes vestigial organs. Both the central nervous and immune system overproduce cells and destroy those that do not form successful synapses or produce usable antibodies.[186]

Destruction of individual cells doesn't stop their DNA from passing to the next generation. All the cells in the body are clones with exactly the same genes. And if a cell's death increases the fitness of the animal as a whole, its sacrifice is "a far, far better thing." It brings only advantages. Life from death.

A dark inheritance

Recall that the first eukaryotic cell was likely formed when a bacterium was engulfed by an archaean. DNA analysis hints the bacterium's suicidal tendencies were transferred to the chimeric new organism at that time. Modern eukaryotic organisms are more or less addicted to programmed Death. They use it to protect both the individual multicelled organism, and the species as a whole.

How did eukaryotes come into possession of Death? Researchers Koonin and Aravind reviewed the genetic evidence. They believe that before the time of the first eukaryote, bacterial parasites were using caspases to control their archaean hosts. A bacterium living inside an archaean might have been using caspase-like chemicals "to kill their host cells once those became unhospitable environments, e.g. because of scarcity of nutrients. Such a mechanism could enable the endosymbionts to efficiently use the corpse of the assassinated host and move to another host. During subsequent evolution, this weapon of aggression might have been appropriated by the host and made into a means of programmed suicide."[187] That host was the first eukaryote. And now it had suicidal tendencies.

So the ultimate beneficiaries of the struggle between bacteria and their phage parasites were the eukaryotes. Bacteria stole Death from the phages and made it their own. The earliest eukaryote, in turn, inherited Death when an archaean took in a Death-wielding bacterial endosymbiote. Death is now a vital part of all eukaryotic life, and complex multicellular organisms (like us) could not exist without it.[188]

Something fishy

We have reviewed some pretty good evidence that, as paradoxical as it may sound, animals use Death to protect their fitness. We have suggested that individual members of sexual species are programmed to die as a means of protecting future generations. Could it also be that programmed senescence, like Death, is another one of nature's tools for fighting off parasites? Are we ourselves destined to enter into "old age" not because our bodies will wear out, but because we are programmed to do so by our own genes?

The programmed aging hypothesis is given support by observations of multiple species that seem to "get old and die" extremely suddenly and quite predictably following reproduction. We have reason to suspect that rapid deterioration and Death are genetically programmed into the life cycles of sexually reproducing eukaryotic animals that mate only once in their lifetimes.

These animals are called "semelparous." That term comes from the Latin words "semel" meaning once, and "pario" meaning beget. (We humans belong to a different class of animals called "iteroparous." The "itero-" here means we beget repeatedly.) Following their single effort at reproduction semelparous animals characteristically fall into rapid deterioration and quickly die, despite their being at the height of their health and vitality just before reproduction. It would be easy to conclude that for these animals Death is simply part of the reproductive program.

The classic example of a semelparous animal is the Pacific salmon. When these salmon swim upriver to spawn they are strong and vigorous. But immediately after reproducing they enter into a state of rapid decline and senescence, and a short time later they are literally falling apart.[189] This simple observation, taken at face value, would seem to imply that – for this animal at least – aging is part of a genetically controlled program.

A similarly impressive example from the plant kingdom is the Madagascar Palm. This semelparous plant flourishes for 100 years, then sexually reproduces. Despite its apparent vigor before seeding, following reproduction it promptly and predictably dies.[190]

The notion of programmed aging doesn't sit well with some people, so this area of research remains a bit contentious. There are some natural barriers that make it hard to objectively evaluate the survival value of programmed death. Some define reproductive success in a strict manner – success means leaving the most offspring. So these researchers doubt that senescence or programmed Death could have any reproductive benefit. Researchers Kirkwood and Melov express these doubts clearly.

> There is, it must be acknowledged, an instinctive attraction to the idea that aging is programmed. Aging is widespread across species and applies universally to all individuals within a species in which it is observed. There is also reproducibility about changes that occur with aging, although close consideration reveals huge inter-individual differences that are hard to reconcile with any strict form of programmed control. But to hold to the idea that aging is programmed, in the face of the evolutionary logic and experimental evidence to the contrary, is as unpromising a scientific stance as to continue to assert that the sun orbits the earth.[191]

Kirkwood and Melov believe the semelparous animal dies after reproduction simply because it exhausts itself in the strenuous one-time drive to reproduce. In their view, the post-reproductive body is vulnerable because it is no longer protected by evolutionary pressures. This is the "disposable soma" theory. Nature, they believe, no longer has a use for the body after its single sexual adventure. After the animal reproduces, Mother Nature essentially says "Your health is no longer my concern." The dramatically sudden craterings of semelparous adults are, they say, "extreme examples of the 'disposable soma.'"[192]

Certainly Death isn't good for the individual who dies. But despite their doubts Kirkwood and Melov acknowledge that Death might benefit close relatives. It can help to pass on genes under certain conditions. According to them, "The best candidates for such programming arise where there is spatial clustering of a population with limited dispersal and where space vacated by the death of an adult is likely to be occupied by its offspring or close kin."[193] In other words, Death makes sense when it brings life to close relatives.

"Space vacated by the death of an adult." Hmm. We have seen that as viruses tighten their grip on the dominant strain in a prokaryotic bloom, that strain faces extinction. When the doomed dominant strain throws in the towel it vacates space for close kin and feeds them with the nutrients that were in their bodies, helping them to survive. So the conditions specified by Kirkwood and Melov are met in such a bloom. Could this logic apply to entire multicelled animals as well?

It most certainly does apply when we're talking about the Death of body cells that make up the organism. In this case the "close relatives" couldn't be any closer – they all have a DNA package that's identical to the one in the sacrificed cell.

Goodbye, Pops

The rationale for programmed Death of cells also seems to apply to breeding populations of animals. Genes must be constantly shuffled to deal with parasites. This being the case, letting old-model, unshuffled, parasite-friendly individuals live too long means allowing parasites to maintain their grip. Especially if those vulnerable individuals continue to breed. So in order to defend the health of the breeding population as a whole it is necessary to usher the old-timers offstage. Goodbye Pops. Farewell Grandpa. And thanks.

What this means is that there is a credible rationale for programmed senescence and Death even in animals that reproduce more than once. Recent thinking seems to be moving toward acknowledging the reality of programmed aging. Researcher V. Skulachev asserts that "the balance between concepts of programmed and non-programmed aging seems to be really shifted to the programmed one." Skulachev believes the tide began to turn as early as 1972 when a major paper[194] concluded that cells of multicellular organisms tend generally to be individually pre-programmed for death.[195] Others acknowledge that the notion of programmed senescence and death is at least logically defensible.[196]

In summary

Chapters 1 and 2 outlined the powerful impact of parasites upon evolution, an influence that has only recently begun to be appreciated. The need to defend against parasites certainly led to the emergence of the immune system. Conversely, parasites' need to get past host resistance led to their ability to degrade their hosts' immune response.

Chapter 3 traced the process by which parasites provoked the evolution of sexual reproduction. Immunity and sex don't generally appear together in the same sentence. Nevertheless, our new perspective places sex under a larger umbrella of immune function.

The things that we normally think of as agents of immunity – antibodies, macrophages, lymphocytes, etc. – arose under selection pressure to protect the individual multicelled organism. Sex, by contrast, serves to protect the breeding population as a whole.

Non-sexual reproduction passes a winning gene package, unchanged, into the next generation. Sexual reproduction also passes genes into the next generation, but never as a complete package. In order to keep parasites at bay, sexual recombination "encrypts" or scrambles the winning packages. With sexual reproduction, the individual winners of the survival game don't get to pass on their complete DNA package. That would give parasites the upper hand. Instead, the prize awarded to winners is more modest – the privilege of adjusting the frequency of genes in the gene pool. This keeps the young of a sexual species moving in the same general direction as their ancestors, but not stepping exactly in their bootprints.

Chapter 4 explored the origin and uses of programmed death. Like sex, it protects the interests of the breeding population. Death of the individual multicellular animal clearly doesn't benefit that animal. But where parasites are a threat, Death does provide benefits. With each new generation, elimination of the previous generation removes gene configurations with which parasites have had time to become familiar. This keeps the parasites off balance.

Chapter Five
The Sunny Side of Senescence

Researcher Skulachev asserts that senescence and Death are under the genetic control of the organism. We get old and die, he says, not because extended life is impossible but because evolution has rejected immortality. But he also makes it clear that programmed senescence need not be a depressing prospect. In fact, he makes a point that is quite positive: "If aging is programmed, it can be retarded, prevented, and perhaps even reversed by treatments interrupting execution of this program, just as we already can interrupt programs of cell death. In other words, programmed aging can be cured like a disease."[197] If the research we just explored holds any water, it is likely that we will in the future be able to slow, adjust, or stop programmed cell death (PCD) by manipulating the body's telomeres, the "end caps" on chromosomes.

Due caution

Before we go further, a few words of caution: When we speak of "curing" senescence or of dodging Death, we are talking about reversing a natural system – one rooted firmly in the history of life on this earth. As we have seen, Death is a natural process that took shape for very good reasons over an unimaginable timespan. It serves a critical purpose. Senescence is cued by a system of elements in delicate dynamic tension. From our births and onward through our lives biochemical factors supporting life are finely balanced against factors pressing us toward Death. Natural selection has judged that balance to be of critical importance.

So we must ask this question: If the purpose of Death is to combat parasites, in manipulating Death will we not be unleashing and empowering those very parasites? Scientists are aware that historically wherever we find disruption of natural senescence we find it in the company of serious disease or bodily dysfunction. Researchers Sanmartin et al state, "Deregulation of apoptosis, the main form of PCD in animals, is associated with diseases such as cancer, autoimmune diseases, and neurodegenerative disorders."[198]

There are in addition significant moral and ethical issues that will come into play if we succeed in delaying senescence, or in pushing Death back just a few years. Who among us will be allowed to push back Death? Will it be only those who can afford it? Is the planet able to support the individuals who would be living longer? With fewer people dying would there have to be fewer children born?

These were the sorts of issues highlighted in the 1974 film *Zardoz*.[199] That movie depicts a future community where death has been vanquished and aging is controlled by the state. Citizens of the community are punished for crimes by making them older artificially. Repeat offenders end up as decrepit, semi-demented seniors confined to nursing homes without even the prospect of release through death. "Uncivilized" non-members who engage in uncontrolled reproduction are regarded as savages. They are hunted down and killed to keep their numbers under control.

To be sure, *Zardoz* is science fiction. But the concerns it highlights are real enough. Seizing control of Death is an audacious act. It requires that we be mindful of what we're getting ourselves into. And we should acknowledge from the outset that we may be unable to fully predict the consequences, or to fully handle the responsibility.

Tickling the telomeres

These cautions noted, there are developments afoot in the field of life extension that are genuinely revolutionary. Many of these do in fact involve manipulation of telomeres. Recall that the human body seems to use the length of a cell's telomeres as a critical age indicator. Programmed senescence of the body is triggered when the telomeres in a critical number of cells get too short. Senescent cells secrete the chemical progerin, which begins the cascade of bodily changes typical of "old age."

Gerontologists have always believed it possible to slow aging under some conditions. But because senescence had been understood in terms of "wearing out," few believed it could actually be reversed.

That conception was strongly jolted on November 28, 2010 by a study reported in the journal *Nature*.[200] The Life Extension Foundation (LEF) summarized it this way:

In this unprecedented study, Harvard-affiliated researchers lengthened telomeres in aging mice and achieved rapid reversal of genetically programmed organ and tissue degeneration caused by short telomeres... The aged mice showed new brain cell growth, restored sexual function and fertility, and regeneration in every tissue examined. These senescent mice dramatically reversed genetically predisposed damage, in particular to the brain and central nervous system, after their telomeres were lengthened.[201]

For this study the researchers used mice that had been genetically altered to have unusually short telomeres. This abnormality produced the same kind of defect as in the progeric children discussed earlier. Their too-short telomeres caused rapid aging. They noted that these mice showed many symptoms typical of senescence – "progressive tissue atrophy, stem cell depletion, organ system failure, and impaired tissue injury responses..."[202]

Jaskelioff et al wanted to see whether lengthening the mice's telomeres could reverse degenerative symptoms of old age, even after the symptoms of senescence were clearly evident.

The results of this early experiment were dramatic. Restoring the telomeres did appear to undo damage that had already been done. In effect, mice that had seemed "old" both visually and clinically "got younger" with 4 weeks of experimental treatment. In the past such changes would have been considered impossible.

Tissues of multiple organ systems regained the characteristics of youth, and several sorts of functioning were more youth-like, including sexual functioning. Importantly, the treated mice lived longer. The researchers conclude that "this unprecedented reversal of age-related decline in the CNS and other organs vital to adult mammalian health justifies exploration of telomere rejuvenation strategies for age-associated diseases, particularly those driven by accumulating genotoxic stress."[203]

Although the Jaskelioff study used mice as subjects, the results are presumed relevant to humans. The researchers point to the parallels between the degeneration exhibited by these altered mice and our own typical symptoms of old age. "Mounting evidence in humans has also provided strong association of limiting telomeres with increased risk of age-associated disease and with onset of tissue atrophy and organ system failure in degenerative diseases."[204] In other words, it seems that telomere loss triggers many of the symptoms of old age, in both mice and humans, and that reversal of such degenerative changes no longer seems impossible.

There is a complication, though. Available data suggests mouse cells may not use telomeres in the same way as do human cells, for "replicative aging" – i.e. aging the body by counting the number of cell divisions. This may make it hard to generalize the Jaskelioff study to humans.

Researchers Gomes et al addressed this problem in a recent study.[205] They examined telomeres and telomerase in cultured cells from over 60 mammalian species to get an idea of how widespread replicative aging is among mammals. Their results confirm that many smaller mammals (including laboratory mice) do not use replicative aging to decide when to die.

Interestingly, the research group was able to establish with some certainty that "ancestral placental mammals had human-like telomeres and repressed telomerase" just as humans do.[206] However, "multiple independent times smaller, short-lived species changed to having longer telomeres and expressing telomerase."[207] In other words, although early in their evolution all mammals were counting their cell divisions as humans do, many smaller mammals abandoned the practice over time. Why? The authors speculate that replicative aging doesn't protect against cancer very well except in larger animals. Interestingly, even though the smaller animals abandoned this particular method, all appear to have remained invested in programmed death. They simply developed other ways of deciding when to die. Senescence and death must be pretty important.

Their data suggests which mammalian models might be appropriate for investigating human cancer and aging. They also observe that primates and Indian Muntjac deer (the oldest known deer, with a fossil record dating from 15 to 35 million years ago) have short human-like telomeres and use replicative aging. So they might be more realistic models than laboratory mice for the study of programmed aging in humans.[208]

In response to the Jaskelioff study, LEF (a major independent funder of research in longevity) announced $2 million funding for an age-reversal study capitalizing on these and related findings. LEF scientists believe that there are even more effective methods for telomere lengthening and aging reversal than were used in the Jaskelioff study. Accordingly they plan a series of additional studies, the first of which will use "several cell rejuvenation mechanisms (including telomere lengthening) in a real world model of aging."

In other words, the LEF studies will determine whether animals that are actually of advanced age (rather than animals genetically engineered to be "old") can in fact be made more youthful. If the results are promising, they plan to extend their findings to humans. An LEF statement concludes that "Based on regenerative medicine technologies being developed right now, we may be only a short time away from making old people young again!"[209]

Hanging on to youth

So it's conceivable that the outlandish prospect of "growing younger" through telomere lengthening could become a reality within our lifetimes. One company is already offering to the public a nutritional supplement (an extract of the herb Astragalus, long used in traditional Chinese medicine) that has documented telomere-lengthening properties.[210] Although this company states they have documented multiple benefits of use through double-blind, placebo-controlled research, they offer no dramatic examples of users "getting younger" even after several years' supplementation.

Adventurous individuals may elect to try this supplement. But those interested in the telomere-lengthening approach to life extension should know there are a number of other straightforward, safe, practical, low-cost steps they can take right now to preserve (and in some cases lengthen) telomeres naturally. Here are some of them:

Calorie restriction

Calorie restriction (eating less, but not to the point of malnutrition) has been shown to reliably lengthen lifespan in a number of species, and may lead to longer life in humans. Calorie restriction regimens typically involve a nutritious diet with calorie intake reduced by 20-50% below normal levels.[211] Researchers Dilova et al write "Calorie restriction (CR) is the most potent regimen known to extend the lifespan in multiple species. CR has also been shown to ameliorate several age-associated disorders in mammals and perhaps humans."[212] Hunt et al add that "CR is the only known intervention reported to extend mammalian lifespan."[213]

Researchers Riesen and Morgan, observing that CR increases lifespan across a broad spectrum of animals, point to "increasing recognition that similar fundamental cellular processes underly aging in all eukaryotes."[214]

Following up on the Jaskelioff mouse study, researchers Vera et al put together an experiment to see whether calorie restriction could enhance the benefits of telomere lengthening. They used the same kind of genetically altered mice as in the earlier study, and came up with some fascinating results. They found CR did enhance the ability of telomerase to extend lifespan and reduce cancer in the altered mice. But quite importantly they found there were similar benefits for a percentage of "wild-type" (unaltered) mice as well. They report that

> [CR] was able to protect from the development of pathologies associated with aging in both [wild type] and [altered] mice, including insulin sensitivity and glucose intolerance, as well as protection from bone loss over time. In addition to protection from age-related pathologies, CR improved other aspects of mouse health such as neuromuscular coordination in both genotypes. Together, these results indicate that the CR protocol used in this study was able to increase the "health span" of both [wild type] and [altered] mice.[215]

They emphasize that their results "demonstrate that CR attenuates telomere erosion associated to aging."[216] In other words, CR delays the negative health consequences of senescence (including risk of cancer) in normal mice.

Again, these results were obtained from a mouse study. Mice are mammals, and are relatively near humans on the evolutionary tree. So the results may apply to humans as well. Further research is needed.

<u>Fish oil supplements</u>

Clinical evidence suggests that omega-3 fish oil supplements are likely to help preserve telomere length. It has long been known that high omega-3 levels protect heart health. One group of researchers examined telomeres in the blood of cardiac patients. Over the five years of the study telomere preservation was found to be greatest in the patients with the highest levels of omega-3 fatty acids, the nutrients found in fish oil. The benefit seemed substantial: The telomeres of patients with the highest level of omega-3s in their systems had only about one-third the shortening of those with the lowest levels of omega-3s. This suggests that regular fish oil supplementation will slow telomere shortening.[217]

A more recent study was even more positive. Researchers recruited 106 healthy sedentary overweight middle-aged and older adults. In this double-blind trial participants received high-dose or low-dose omega-3 supplements, or a placebo.

After four months of supplementation, results showed a health-protective effect for the omega-3 groups. Authors found that "a decreased ratio of omega-6 [to] omega-3 was associated with longer telomeres, which suggested that lower omega-6 [to] omega-3 ratios 'can impact cell aging.'" Supplementation appeared to significantly decrease inflammation in subjects. "This finding strongly suggests that inflammation is what's driving the changes in the telomeres," said lead author J. Kiecolt-Glaser.[218]

Healthy lifestyle

Healthy lifestyle is quite clearly associated with preservation of telomere length. Researchers Mirabello et al examined lifestyle variables as they related to preservation of telomeres in humans. "Healthy lifestyle factors (i.e., lower BMI [body mass index], more exercise, tobacco abstinence, diets high in fruit and vegetables) tended to be associated with greater telomere length."[219][220]

Another group of researchers (Tiainen et al) studied the relationship between dietary factors and leukocyte telomere length (LTL) in an elderly population. They found that lower total fat and saturated fat intake were related to greater telomere length in men. For women, vegetable intake was positively associated with longer telomeres. The authors conclude that in general their results "support the hypothesis that fat and vegetable intakes were associated with LTL."[221]

Vitamin D and multivitamin supplements

Vitamin D levels as well as multivitamin intake were found to be important for telomere preservation. One set of researchers measured vitamin D levels in over 2000 female twins aged 18 through 79 years. Their findings suggest that "higher vitamin D concentrations, which are easily modifiable through nutritional supplementation, are associated with longer LTL [leukocyte telomere length], which underscores the potentially beneficial effects of this hormone on aging and age-related diseases."[222]

A second set of researchers looked at the connection between multivitamin use and telomere length in women. They report that

> *After age and other potential confounders were adjusted for, multivitamin use was associated with longer telomeres. Compared with nonusers, the relative telomere length of leukocyte DNA was on average 5.1% longer among daily*

multivitamin users... In the analysis of micronutrients, higher intakes of vitamins C and E from foods were each associated with longer telomeres, even after adjustment for multivitamin use. Furthermore, intakes of both nutrients were associated with telomere length among women who did not take multivitamins.[223]

Note that levels of vitamins C and E, both antioxidant vitamins, seemed especially important in preserving telomere length.

Overall, the simple approaches outlined above seem well worth further consideration for anyone who interested in improving health and extending lifespan by protecting their telomeres.

Final comments

Whew! That was a lot of information to cram into a small book. I trust the picture I have painted illustrates just how much parasites have had to do with who we are, and how greatly such creatures have influenced the way we function as human beings.

The pernicious influence of parasites prompted the evolution of our immunity. Intense and never-ending selection pressures can shape intricate biological processes. Our elaborately crafted immune functions are among the finest examples of this intricacy. Central aims of this book have been to clarify that the often-bewildering mechanisms of sex and programmed death also originated as responses to parasites, and to explain how they fit beneath the larger umbrella of immunity.

It certainly isn't necessary to harbor warm feelings toward destructive little creatures that have exploited and sickened us since the dawn of time. On the other hand, it's clear that we wouldn't be what we are today without parasites' influence. I have striven to underline just how powerful that influence has been. I hope I have done the job well enough to make the presentation enjoyable.

Series foreword

All of us can focus our attention. We have no trouble concentrating on particular things in order to bring them to the full light of our awareness. Likewise, we can easily exclude things from our focus, and so move them to the periphery of our awareness. We do these things easily, fluidly, so naturally that we take the process for granted. But where did these abilities come from? How do they relate to consciousness in general? Why is focal attention there at all?

Dead Sexy, the first book in the "Viral Schemes Meet Puppet Dreams" series, examines evidence that several of our complex biological features evolved as defenses against parasites. What will happen if we shine the light of these same principles on psychological phenomena? Does focal attention also have something to do with parasites? Well, that's a question worth asking, and one worth answering. It's the central question addressed by the "Viral Schemes" series.

If we want credible answers to such questions, it certainly makes sense to approach them from an evolutionary perspective. Our ability to pay close attention must have developed because it gave our ancestors an adaptive advantage, an improvement in their ability to survive.

Awareness and attention are mental and behavioral phenomena. So if focal attention is in some way a defense against parasites, we need not expect the parasites in question to be the biological kind – the ones that cause diseases of the body. These would have to be psychological and behavioral parasites, diseases of thought and action.

There are in fact a multitude of malignant patterns in behavior and culture that could be considered behavioral parasites. These are destructive, self-reproducing behaviors that evade our control. Sometimes they evade our awareness as well.

Intuitively, it seems that we come to grips with our own problem behavior only when we are able to "snap" to what we are doing – to bring it to the full light of our awareness. It's clear that in this way our attention protects us from our problem behavior.

Is it possible that focal awareness performs a service for our behavior and cognition paralleling the service our immune system performs for our physical bodies? Is focal awareness simply behavioral immunity? It would be easy to dismiss such questions as absurd. But are they? The five books in this series argue that thwarting behavioral parasites is a central function of focal attention.

The books address these questions systematically. The central argument reflects my own views, not mainstream thought. As far as I can tell, the interpretation of attention as behavioral immunity is an original idea. At least, I haven't run across it in my own readings.

Because it is unfamiliar, I have no doubt that this line of thinking will seem strange to many readers. But give it some time. You may find that, like a parasite, it "grows on you."

Thomas Whitehead

Series overview

Dead Sexy is the first book of five in the "Viral Schemes Meet Puppet Dreams" series. The books are about behavioral viruses – destructive, self-reproducing patterns in our behavior. The aim is to present these ideas:

a) Behavioral viruses are real and common.
b) Behavioral viruses have a profound impact on our day-to-day functioning.
c) A major function of focal awareness is to protect against behavioral viruses.

Here are synopses of each of the five books.

Book One: *Dead Sexy*
How an arms race with parasites gave us sexuality and mortality.

Parasites are everywhere in the animal kingdom. We have underestimated their influence on evolution. We now know that the most significant Darwinian struggle for survival is often that between an organism and its parasites. The selection pressures engendered by this struggle have created some of the most mysterious characteristics of life. Immunity is an impressively complex weapon that evolved in the arms race with biological parasites. Sex and programmed death are additional examples of extreme adaptations for coping with parasites, adaptations that also took shape under the pressure of natural selection. Immune functions, sex, and death share a common purpose. They all fit under the grand umbrella of parasite resistance. But they differ in their areas of focus. While our immune functions enhance our individual fitness, sex and programmed death enhance population fitness. These examples make it easier to understand that our capacity for attention may be the evolutionary fruit of a similar struggle for survival, an ongoing struggle that is taking place within the arena of human behavior.

Book Two: *Puppet Dreams*
The not-so-real world of our personal experience.

We perceive in rich detail the world around us. But the richness doesn't flow directly from the world outside us. Rather, our experience is tied most closely to a detailed internal model of our surroundings. Each of us develops an internal representation of the external world. This model can be called "puppet-like" because, like a puppet, it is an animated caricature of the world it represents. The model is in constant motion, kept in meticulous register with incoming stimuli through evolutionarily-tuned boundary keeping mechanisms. It works this way: Changes in incoming stimuli automatically change the model. Changes in the model automatically change our experience. And changes in our experience impact our behavior. This system for internally representing the external world is an astounding evolutionary development. It enables humans to act with precision in those areas most critical to our survival, greatly enhancing our Darwinian fitness. Even so, the system contains an Achilles heel. Behavioral viruses find ways to directly manipulate the model, giving them enough control over our behavior to reproduce themselves.

Book Three: *Fish in Search of Water*
Why we are blind to the behavioral viruses all around us.

A behavioral virus is a repetitive, self-reproducing pattern that can include action, perception, and thought. We might imagine that such viruses are rare, but they are not. Examination shows that many well-known behavioral patterns fit formal criteria for behavioral viruses. We are in fact immersed in them. One reason we remain unaware of them is that we are used making sense of them in other ways. But there is another, darker reason as well. Each behavioral virus is surrounded by a curious, unnatural distortion of awareness, a distortion without which the pattern would not persist. This systematic interference cloaks the pattern, permitting it to remain unnoticed while in plain sight. This induced blindness is a source of much human dysfunction.

Book Four: *Imaging the Invisible*
How the need for behavioral immunity sharpened awareness.

Parasites and host organisms are locked in a never-ending arms race. Host organisms evolve immune functions. For their part, parasites evolve ways to thwart host immunity. An analogy between biological and behavioral viruses leads to two expectations: a) that the emergence of behavioral viruses would have been accompanied by the parallel development of behavioral immune functions in the host; and b) that behavioral viruses would of necessity develop means of suppressing those immune functions. We expect, in other words, to find an arms race in progress within the behavioral arena. Each behavioral virus is accompanied by a distortion of awareness. And we must overcome the distortion in order to ensure that our behavior serves our own interests. There is evidence that focal attention plays a critical role in dealing with behavioral viruses. We should consider, then, whether focal attention can be best understood as behavioral immunity.

Book Five: *Return to Eden*
Coping with behavioral parasites.

How can we best deal with behavioral parasites? The analogy between biological and behavioral parasites suggests answers. We know quite a bit about thwarting biological parasites. We might reasonably question whether principles of prevention and treatment already validated within the biological realm might have useful analogues in the cognitive/behavioral realm. Looking closely, we find that practitioners in the behavioral arena already use many analogous methods to deal with persistent behavioral problems.

Coming Attractions

Dead Sexy is the first of five books in the "Viral Schemes Meet Puppet Dreams" series. These books introduce behavioral viruses as self-reproducing, repetitive behaviors. And they present focal attention as a means of fighting against them – immunity at the behavioral level.

Book Two is entitled *Puppet Dreams: The not-so-real world of our personal experience.* It deals with our subjective experience of the world around us. It provides a way to understand how our sense of our environment can be distorted and manipulated by behavioral viruses.

Puppet Dreams highlights:

Our experience of the world isn't tied directly to that world. Instead, what we experience comes from an internal model of our surroundings. Evolution has created within human beings a most sophisticated way of interacting with our environments. Somewhere inside each of us is a detailed model of the outside world. We can call this model "puppet-like" because, just like a puppet, it is an animated caricature. It is a caricature of our environment. The model isn't nearly as complex as the world, but it does represent things important for our survival. A mysterious part of us keeps the model in close register with events outside. This registration process works below our conscious awareness, as automatically as our eyeblinks or our breathing. Intricate mechanisms evolved just for this purpose keep the puppet dancing smartly to the tune of incoming stimuli.

The state of the puppet at any given time determines our subjective experience. What we experience is like a waking dream. It's a special sort of dream that is tightly controlled and very realistic – not at all dreamlike. The purpose of the dream is to represent the world around us. But the dream is certainly not that world. It is a map, not the territory the map represents. Many things in the world are not in our dream, and many things in our dream are not in the world. The dream directly reflects the puppet; it does not directly reflect the environment.

Yet we act just as if the dream were reality itself. We depend upon it to guide our behavior in the real world. We rely upon it heavily and completely. In truth we have no choice, because we have no other guide. It just isn't possible for us to experience the world directly.

This remarkable puppet/dream system enables us to act with precision in areas critical to our survival. Our internal map is priceless because it empowers us to navigate efficiently through our environment. Some other large-brained animals have similar capabilities. But this kind of system is most highly developed in humans. It is our most distinctive feature.

Despite the adaptive advantages it confers, the puppet/dream system introduces vulnerabilities. There is a direct chain leading from puppet to dream to behavior. This chain gives viral processes a way to influence us. As explained in *Puppet Dreams*, behavioral viruses are able to sway the puppet in a variety of ways. When they change our world model they change our awareness, and so change our behavior. In this way they gain enough control to arrange their own reproduction. This is good for the virus, but not so good for us. Much human nonsense and misery can be attributed to the malignant influence of behavioral viruses.

The damage caused by biological parasites provoked the evolution of biological immunity. The dysfunction wreaked by behavioral parasites has provoked the evolution of "behavioral immunity." We don't call it immunity, though. We call it attention. It is focused attention that gives us our best shot at escaping the grip of non-productive, parasitic behavioral patterns.

If you have enjoyed *Dead Sexy*, please consider reading *Puppet Dreams* as well. It is scheduled for release in July or August of 2013 (more or less).

Acknowledgments

As I have learned so very well over the past few months, it's hard to do anything without the support of others. This book is a case in point.

From the beginning of this project my stepson Glenn served as a one-man test audience while I pitched ideas. His input helped me clarify what I was trying to say. He critically reviewed later drafts of the book too. Thank you Glenn.

From start to finish my wife Glenda was patient and supportive. She listened to me talk for hours about the wonders of eukaryotes (and other topics she had heard way too much about already). Her kind indulgence helped me to hold the project together during the much-too-long time it took to finish it. Thank you Glenda!

After reading an earlier draft my sister Frannie gave me some critical feedback that made the final product a whole lot better. Thanks Frannie!

My colleague Denise reviewed and commented on a version of the book. She offered some excellent ideas about making the text more engaging and marketable. Thank you Denise!

My brother Bill volunteered to read a later draft of the book, and offered a delightfully uplifting review that kept me moving to the finish line. Thank you Bill!

Finally, thanks to all you thoughtful people who read *Dead Sexy*. Where would we be without you?

NOTES AND REFERENCES

[1] Seilacher A, Reif W, and Wenk P. The parasite connection in ecosystems and macroevolution. *Naturwissenschaften* 2007, 94, 155-169. Page 155.

[2] Zimmer C. *Parasite Rex: Inside the bizarre world of nature's most dangerous creatures.* Touchstone Books, 2000. Page xxi.

[3] Parasites are fascinating, and (as will soon become clear) vastly important. Dead Sexy is the first in a series about attention and awareness. What do parasites have to do with awareness? Plenty! I argue that our capacity to pay close attention to things evolved as a way of defending against parasites. Not the familiar biological parasites, mind you. Attention is a remedy for behavioral parasites – persistent, self-reproducing patterns of behavior that exploit our ability to respond to our environment in a complex and flexible way. The intent of this series is to give this idea serious consideration. To understand how parasites could so dramatically impact our psychology, we need an appreciation of the way they have shaped our biology. Host animals, in order to cope with their parasites, have had to adopt extreme measures. This first book details just how far they have had to go. Casting light now on the complexity of these adaptations will help to explain later how similar pressures within the behavioral arena might have led to the evolution of focal awareness.

[4] Maybe we regard "diseases" as a special category because we feel uneasy about harboring parasites, and are less threatened when we're told we have a disease. But disease organisms are really parasites.

[5] Seilacher et al 2000. Page 159.

[6] Davies N. Cuckoos. Article in *Current Biology*, 2007, 17, 10, R346-R349. Page R346.

[7] With a giant monster in their nest in plain view you'd think the host parents would notice something was wrong. We might be tempted to distance ourselves from the plight of these bird-brained victims. We could label their short-sightedness pathetic, or tragic, or even darkly comical. But we might do better to learn from their example. The truth is that we humans err in exactly the same way, for the same reasons, and with much the same result. We are blind to our errors just as these parents are. No creature has the power to see the world as it is. All animals (us included) use cues to build interpretations – rough sketches – of what's happening. Our interpretations come from inside us, and are limited to what we are already prepared to see. The cuckoo chick is tuned by natural selection to give just the right cues to create a false impression. Because the cues are right, the parents perceive the alien intruder as their offspring, and their chick-rearing as being on track. It looks right – to them. As human onlookers we can spot their mistake because our powers of perception and comprehension are

(somewhat) greater.

[8] Davies 2007. Page R347.

[9] Davies 2007. Page R348.

[10] Davies 2007. Page R346.

[11] Bickford C, Kolb T, and Geils B. Host physiological condition regulates parasitic plant performance: Arceutholbium vaginatum subsp. Cryptopodum on Pinus ponderosa. *Oecologia*, 2005, 146, 179-189. Page 179.

[12] Strauss J and Strauss E. *Viruses and Human Disease*, Academic Press, 2002. Page 4.

[13] Strauss J and Strauss E 2002. Page 1.

[14] Zimmer 2000. Page 14.

[15] Darwin C. The role of a hypothetic creator is compatible with the evolution of the biological species, 1842 (a 10-page pencil-written early sketch of theory).

[16] Quoted in Desmond A and Moore J. *Darwin: The life of a tormented evolutionist*. 1994, Norton Books. Page 293.

[17] Also quoted in Patton M. *Science, Politics and Business in the work of Sir John Lubbock: A man of Universal Mind*. 2007, Ashgate Publishing, Page 35.

[18] Seilacher et al 2007. Page 159.

[19] Dawkins R. *The Extended Phenotype: The long reach of the gene*. Oxford University Press, USA, Revised edition (August 5, 1999)

[20] The science fiction film Invasion of the Body Snatchers was directed by Don Siegel and starred Kevin McCarthy and Dana Wynter. It was released in 1956 and distributed by Allied Artists Pictures Corporation. In the story aliens invade a small town, waiting until people fall asleep to replace them with alien duplicates that look just like them.

[21] The science fiction/horror film *Alien* was directed by Ridley Scott, with Sigourney Weaver in the starring role of Ripley. It was released in 1979 and distributed by 20th Century Fox. The plot revolves around an extremely dangerous and intelligent parasitic alien lifeform that stalks and kills the crew of Ripley's commercial spaceship one by one.

[22] Fenton A and Rands S. The impact of parasite manipulation and predator foraging behavior on predator-prey communities. *Ecology*, 2006, 87(11), 2832-2841

[23] Seilacher et al 2007. Page 160.

[24] Seilacher et al 2007. Page 160.

[25] Lafferty K, Hechinger R, Shaw J, Whitney K, and Kuris A. Food webs and parasites in a salt marsh ecosystem. Pages 119-134 in Collinge S and Ray C, editors. *Disease ecology: Community structure and pathogen dynamics*. 2005, Oxford University Press. Page 129.

[26] Seilacher et al 2007. Page 160.

[27] Seilacher et al 2007. Page 160.

[28] The principle of opportunism is important for understanding biological

parasites, because it explains how simple organisms can produce complex host behavior. This principle is just as important when we consider parasites operating in a different arena – that of psychology and behavior. Within this arena very simple patterns of self reproducing behavior exploit the complex capabilities of their hosts – human beings like us. In this way complex behavior can be engendered by a very simple behavioral virus.

[29] Fenton and Rands, Pages 2832-2833.

[30] Henriquez S, Brett R, Alexander J, Pratt J, Roberts C. Neuropsychiatric disease and Toxoplasma gondii infection. *Neuroimmunomodulation.* 2009;16, 2, 122-33. Page 122.

[31] Skallova A et al. Decreased levels of novelty seeking in blood donors infected with Toxoplasma. *Neurological Endocrinology Letters*, 2005, 5, 480-486. Page 480.

[32] Zimmer 2000. Page 195.

[33] Ferguson DJ, Hutchison WM. The host-parasite relationship of Toxoplasma gondii in the brains of chronically infected mice. *Virchows Archiv A, Pathological Anatomy and Histopatholog*y, 1987, 411, 1, 39-43. Page 39.

[34] Drisdelle R. Toxoplasma Gondii and Behavior. Online article, Suite101.com, April 17, 2007.

[35] Boulter N. Alley cats and sex kittens. *Australasian Science*, January/February 2007, 35-37. Online copy retrieved 3/10/2013 from http://www.control.com.au/bi2007/281parasites.pdf

[36] Quoted in Anonymous. Parasite makes men dumb, women sexy. *The Sydney Morning Herald*, December 26, 2006. Retrieved 3/10/2013 from http://www.smh.com.au/articles/2006/12/26/1166895290973.html?from=to p5

[37] Flegr et al. Induction of changes in human behavior by the parasite protozoan Toxoplasma Gondii. *Parasitology*, 1996, 113, 49-54.

[38] Flegr J et al. Increased incidence of traffic accidents in toxoplasma-infected military drivers and protective RhD molecule revealed by a large-scale prospective cohort study. *Biomed Central Infectious Diseases*, 2009, 9, 72.

[39] Skallova A et al 2005. Page 480.

[40] Henriquez et al 2009. Page 122.

[41] Henriquez et al 2009. Page 122.

[42] Fels D, Lee V, and Ebert D. The impact of microparasites on the vertical distribution of Daphnia magna. *Archives of Hydrobiology*, 2004, 161, 65–80.

[43] Bhatt G. Parasite Influence on Host Behavior - Part 1. Online article dated 7/7/07, Article ID 5853, collected 2/12/13 from http://www.boloji.com.

[44] Thomas F, Schmidt-Rhaesa A, Martin G, Manu C, Durand P, and Renaud F. Do hairworms (Nematomorpha) manipulate the water seeking behaviour

of their terrestrial hosts? *Journal of Evolutionary Biology*, 2002, 15, 3, 356–361.

[45] Von Bertalanffy L. *General System Theory: Foundations, development, applications* (Revised Edition). George Braziller Inc. 1969. Page 41.

[46] Why talk about boundary keeping here? The answer is that there is a continuity between the simplest processes by which life is maintained and the most complex functions of intelligent behavior. The behavior of all living things – including the brain-based, intelligent behavior of higher animals – serves a common purpose. And that purpose is increasing chances of survival. "Intelligent" behavior can be produced by evolutionary processes – with no brain involved – or it can be produced by conscious consideration and deliberation. Brain-based intelligence is more flexible, much faster, far broader in scope, and in many cases more effective than the behavior of primitive cells. But in the final analysis, it's all a manifestation of the same basic drive to survive. Here I emphasize the progression from boundary keeping to intelligence in order to create a foundation for a later discussion of awareness. I hope it will become apparent that boundary keeping processes are central to focal attention, just as they are to basic life processes.

[47] Damasio A. *The Feeling of What Happens: Body and emotion in the making of consciousness*. Mariner Books, 2000. Pages 138-139.

[48] Vellai T and Vida G. The origin of eukaryotes: the difference between prokaryotic and eukaryotic cells. *Proceedings of the Royal Society: Biological Sciences*, 1999, 266, 1428, 1571-1577. Page 1574.

[49] Kochin B, Bull J, and Antia R. Parasite evolution and life history theory. *PLoS Biology*, 2010, 8, 10, 1-4. Page 1.

[50] My description of the immune system is vastly oversimplified – just detailed enough to provide an overview of this amazing set of defensive mechanisms. The idea is to explain what the system is and why it came into being. This will let us compare the functions of sex and programmed death a little later in the book.

[51] A central theme of the "Viral Schemes" series is that focal attention provides immune functions protecting the complex behavior of organisms such as ourselves – organisms whose behavior is guided by an internal representation of their external environment. Several features of this "behavioral immunity" closely mirror features of biological immunity. So it is worthwhile to examine those features early in the series.

[52] Again, this description of the innate immune system is seriously oversimplified, and doesn't do it justice. Innate immunity is superbly designed by evolution, and does its job so smoothly and effectively that we almost never notice its operation.

[53] If we tried to recognize our friends the same way, would it work? Well, yes and no. If we recognized Gloria's nose we might get an image of her standing in front of us. But we would have to admit that our image of Gloria

doesn't really come from "out there" in the world. It would be an image pulled from inside ourselves, based on our experience with Gloria and our assuming her nose is connected to the rest of her. Most of the time we'd be right. But sometimes we'd be wrong. In fact our perception does seem to work something like "recognition-by-trigger," and for this reason our perceptions sometimes don't match reality.

[54] We're keeping the discussion of adaptive immunity simple because our purpose is to illustrate the intricacy of immunity, and to emphasize that its impressive level of refinement is a direct consequence of the arms race with parasites.

[55] That is, almost all are identical. But the system is set up to produce mutations at a high rate – about one mutation in every couple of thousand units. The mutations produce subtle variations in the shape of the "grabbing" or antibody end of the lymphocyte. Presumably over time these variations in the basic shape further refine the antibody to match the invader even more exactly.

[56] Nossal G. Life, death and the immune system. *Scientific American*, September, 1993, 53-62. Page 54.

[57] Marrack P and Kappler J. How the immune system recognizes the body. *Scientific American*, September, 1993, 81-89. Page 81.

[58] The body routinely creates organs and systems of cells by starting out with a large number and "sculpting" them by killing off the ones it doesn't need. This may seem harsh. But as will be explained presently the interests of the killed cells are served by the process.

[59] Marrack P 1993. Page 81.

[60] Marrack P 1993. Page 81.

[61] Development of a similar self/other distinction can also be observed within the realm of human behavior. Our individual system of personal boundaries is typically developed within our family of origin. Women who have grown up in a home atmosphere of domestic violence are far more likely to accept domestic violence in their own intimate relationships than are women who have no such family background. Seemingly a first incident of violence in an intimate relationship does not trigger an impulse to end the relationship, as it generally would in a woman from a family of origin where no domestic violence occurred. Such boundary systems are discussed later in the series.

[62] Paul W. Infectious diseases and the immune system. *Scientific American*, September, 1993, 91-97. Page 96.

[63] Nossal G 1993. Page 55.

[64] Paul W 1993. Page 96.

[65] Paul W 1993. Page 97.

[66] We can also see this sort of abandonment of immunity at the level of behavior and culture. Sometimes our efforts to control a pathological pattern of behavior seem to cause more damage than the pattern itself. At

that point we often inactivate what passes for immunity at the level of behavioral function. The pathological pattern then runs rampant.

[67] Finlay B, and McFadden G. Anti-immunology: evasion of the host immune system by bacterial and viral pathogens. *Cell*, 2006, 124, 4, 767–782.

[68] Hornef M, Wick M, Rhen M, and Normark S. Bacterial strategies for overcoming host innate and adaptive immune responses. *Nature Immunology*, 2002, 3, 11, 1033-1040.

[69] Farrell H, Degli-Esposti, and Davis-Poynter N. Cytomegalovirus evasion of natural killer cell responses. *Immunology Reviews*, 1999, 168, 187–197.

[70] Murphy P. "Molecular piracy of chemokine receptors by herpesviruses". *Infectious Agents and Disease*, 1994, 3, 137–164.

[71] Clarity about this "screw-it-up-or-die" principle will be critical later, when we evaluate evidence that a different kind of viral process – the behavioral virus – is more common than we realize. This viral characteristic will become obvious as we study the impact of behavioral viruses on the cognitive/behavioral capacities that provide immune functions there.

[72] Borrow P, Evans C, and Oldstone M. Virus-induced immunosuppression: Immune system-mediated destruction of virus-infected dendritic cells results in generalized immune suppression. *Journal of Virology*, 1995, 1059-1070. Page 1059.

[73] Humans and other complex multicellular animals use a similar system for producing behavior. Each of us has the ability to engage in thousands of particular behaviors on demand – behaviors of vastly different sorts. Some form of each is stored within us as a potential. These potential behaviors are turned into actual behaviors by means of a system so familiar to us that we generally fail to appreciate its complexity.

[74] Rajnik M. Rhinovirus infection. Medscape Reference. Webpage http://emedicine.medscape.com/article/227820-overview#a0104 retrieved 3/28/13.

[75] Later in the series we will come back to this point, when we're talking about behavioral parasites. The idea is that certain complex patterns of human behavior – behaviors that seem pretty sophisticated – are the creations of relatively simple behavioral viruses.

[76] This series is billed as being about awareness. The interpretation of sex and death as defenses is interesting. But the reader may rightly wonder what value this interpretation brings to the consideration of attention and awareness. The answer is that sex and death serve as excellent examples of the extremes to which animals must go in order to stay alive in a world filled with parasites. Looking closely at these complex adaptations clarifies their surprising relationship to immunity. This sets the stage for a later assertion that focal awareness is also an immune function, an adaptation for coping with another kind of parasite – the behavioral virus.

[77] Zimmer C. On the origin of eukaryotes. *Science*, 2009, 325, 5941, 666–

668.

[78] The word "cytoplasm" is a scientific term cobbled from the Greek words "kytos" meaning container and "plastos" meaning molded. The cytoplasm is like the jello in a mold, you see.

[79] Waggonner, B. Interpreting the earliest metazoan fossils: What can we learn? *American Zoology*, 1998, 38, 975-982.

[80] Knoll A. The Multiple Origins of Complex Multicellularity. *Annual Review of Earth and Planetary Sciences*, 2011, 39, 217-239. Page 218.

[81] Knoll A 2011. Page 218.

[82] Phoenix C. Cellular differentiation as a candidate "new technology" for the Cambrian Explosion. *Journal of Evolution and Technology*, 2009, 20, 2, 43-48.

[83] Lane N. *Life Ascending: The 10 great inventions of evolution.* Norton books, 2009. Page 90.

[84] Lane N 2009. Page 94.

[85] National Institutes of Health. NIH Human microbiome project defines normal bacterial makeup of the body. *NIH News*, Wednesday, June 13, 2012. Http://www.nih.gov/news/health/jun2012/nhgri-13.htm

[86] Levine A. Viruses. *Scientific American Library,* 1992. Page 18.

[87] We will later argue that a similar capacity for emulation within the behavioral arena is one of our most basic characteristics as human beings. Complex vertebrates are able to act in very different "modes" depending on circumstances and need. This is most obvious when we consider our own acts. At different times we behave like different animals. At times we are lambs, at times we are lions. Sometimes we are even parasites. Our flexibility permits us to temporarily assume a sequence of very different (sometimes incompatible) learned behavioral modes. We transform ourselves into a succession of what we might whimsically call virtual animals. Unfortunately, behavioral viruses have taken full advantage of this particularly human capacity.

[88] Szathmary E, and Smith J. The major evolutionary transitions. *Nature*, 1995, 374, 227-232.

[89] Fuerst J. Beyond Prokaryotes and Eukaryotes : Planctomycetes and Cell Organization. *Nature Education*, 2010, 3, 9, 44.

[90] Lane N. 2009. Page 102.

[91] Kelly S, Wickstead B, and Gull K. Archaeal phylogenomics provides evidence in support of a methanogenic origin of the Archaea and a thaumarchaeal origin for the eukaryotes. *Proceedings of the Royal Society Biological Sciences*, 2011, 278, 1009-1018. Page 1016.

[92] Lane N 2009. Page 89.

[93] Fuerst J 2010. Page 44.

[94] Koonin E and Aravind L. Origin and evolution of eukaryotic apoptosis: the bacterial connection. *Cell Death and Differentiation*, 2002, 9, 4, 394-404. Page 402.

Page 131

[95] Frade J and Michaelidis T. Origin of eukaryotic programmed cell death: a consequence of aerobic metabolism? *Bioessays*, 1997, 19, 827-832.

[96] Sigismund S et al Endocytosis and signaling: Cell logistics shape the eukaryotic cell plan. *Physiological Reviews*, 2012, 92, 273-366. Page 276.

[97] Sigismund S et al 2012. Page 274.

[98] Sigismund S et al 2012. Page 342.

[99] Sigismund S et al 2012. Page 273.

[100] Sigismund S et al 2012. Page 273.

[101] Mercer J, Schelhaas M, and Helenius A. Virus entry by endocytosis. *Annual Review of Biochemistry*, 2010, 79, 803-33. Page 803.

[102] Mercer J et al 2010. Page 805.

[103] Mercer J et al 2010. Page 805.

[104] Futuyma D. *Evolution*, second edition. Sinauer Associates, Inc, 2009. Page 350.

[105] Lane N 2009. Page 139.

[106] Neiman M, Hehman G, Miller J, Logsdon J, and Taylor D. Accelerated mutation accumulation in asexual lineages of a freshwater snail. *Molecular Biology and Evolution*, 2009, 27, 4, 954.

[107] Williams G. *Sex and Evolution*. 1975. Princeton University Press.

[108] Quoted in Cartwright J. *Evolution and Human Behavior: Darwinian perspectives on human nature*. 2000, MIT Press. Page 97.

[109] Ridley M. *The Red Queen: Sex and the Evolution of Human Nature*, 2003, Harper Perennial.

[110] Ridley M 2003. Page 72.

[111] Morgan A and Buckling A. Relative number of generations of hosts and parasites does not influence parasite local adaptation in coevolving populations of bacteria and phages. *Journal of Evolutionary Biology*, 2006, 19, 6, 1956-1963. Page 1956.

[112] Ridley M 2003. Page 72.

[113] Lane, 2009. Page 135.

[114] King K, Jokela J, and Lively C. Parasites, sex, and clonal diversity in natural snail populations. *Evolution*, 2011, 65, 5, 1474-1481.

[115] Zuk M. *Sex on Six Legs: Lessons on Life, Love, and Language from the Insect World*. 2011. Houghton Mifflin Harcourt Publishing. Page 171.

[116] See also Law J and Regnier F. Pheromones. *Annual Review of Biochemistry*, 1971, 40, 1, 533-545.

[117] For example, the sea cucumber (Holothuroidea) can breed both sexually and asexually. When breeding sexually "The animals release both eggs and sperm into the water and fertilization occurs when they meet. There must be many individuals in a sea cucumber population for this reproductive method to be successful." Anonymous. *National Geographic*. Webpage http://animals.nationalgeographic.com/animals/invertebrates/sea-cucumber/ retrieved 3/20/13.

[118] See also Regal P. Pollination by wind and animals: Ecology of

geographic patterns. *Annual Review of Ecological Systems*, 1982, 13, 497-524. Page 497.

[119] Glover B. *Understanding flowers and flowering: an integrated approach.* 2007, Oxford University Press, Page 127.

[120] See description and hear mating sounds on The University of Michigan Museum of Zoology, Insect Division. Periodical Cicada Page. Webpage http://insects.ummz.lsa.umich.edu/fauna/michigan_cicadas/Periodical/#Magicicadabehavior. Retrieved 3/20/13.

[121] Stanger-Hall K, Lloyd J, Hillis D. Phylogeny of North American fireflies (Coleoptera: Lampyridae): implications for the evolution of light signals. *Molecular Phylogenetics and Evolution*, 2007, 45, 1, 33–49.

[122] Anonymous. Male Birds' Ability To Learn Song Affects Female Mating Response. *Science Daily*, September 11, 2002. Webpage retrieved 3/20/13.

[123] The sexual appeal of bird songs is often modified by the visual displays and activities of males while they are singing. See O'Loghlen A and Rothstein S. It's not just the song: male visual displays enhance female sexual responses to song and brown-headed cowbirds. *The Condor*, 2010, 112, 3, 615-621. Page 615.

[124] In the 1999 satirical movie *Austin Powers: the Spy who Shagged Me* the character Fat Bastard (a brilliant creation of actor Mike Myers) sensuously proclaims "I'm dead sexy." The remark tickled viewers, because dead sexy is exactly what Fat Bastard is not. He is grossly overweight, crude in his speech and habits, and alarmingly lacking in personal hygiene. The phrase "dead sexy" is likely an offshoot of the earlier "drop dead gorgeous," meaning stunningly beautiful. The Phrase Finder website provides this information about the term's evolution: "'Drop-dead gorgeous' seems to have been with us since just 1985. A piece about Michelle Pfeiffer in Time in February of that year says: 'Trim, smart and drop-dead gorgeous, Pfeiffer has been nibbling at stardom since her stints in Grease II and Scarface.' The phrase struck a chord and there are many references to it in newspapers and journals from very soon after that. It didn't arrive out of the blue. The term 'drop dead', meaning excellent had been around since at least 1962. In The New York Herald-Tribune, January 1962, we have: 'Fashions from Florence not drop-dead. For almost the first time in history Simonetta failed to deliver an absolutely drop-dead collection.' It got picked up as an intensifier for various things, as here from the Washington Post, July 1980: 'For drop dead chic food, Harborplace has a sushi and tempura bar.'" Quotation from The Phrase Finder website, http://www.phrases.org.uk, as retrieved 3/6/2013.

[125] Shakespeare W. *Romeo and Juliet.* Act 2, Scene 2.

[126] Interpretation as detailed by Pressley J. Shakespeare Resource Center. Romeo and Juliet: "But, soft! what light through yonder window breaks...." http://www.bardweb.net/content/readings/romeo/lines.html accessed 3-11-2013.

[127] James Joyce, *Ulysses*. Closing lines.
[128] Excerpt from Folsom, T. *Escorts and Thieves*, Kindle Edition, 2012-12-21, Kindle Locations 320-325.
[129] Henderson J and Anglin J. Facial attractiveness predicts longevity. *Evolution and Human Behavior*, 2003, 24, 351-356. Page 351.
[130] In the next chapter readers will be presented with an argument that senescence and death serve a vital purpose for a population of breeding animals. Programmed death, it is argued, removes combinations of gene alleles with which parasites have had an opportunity to become familiar. With the coming of each new generation, programmed death removes the older generation from the playing board. This prevents them from contaminating the younger generations with genes whose code has already been broken. This could be called the "benevolent reaper" theory. It would predict that sexually active adults should experience a gradient of age-related sexual attraction. Younger adults should be less attracted to potential partners as their age increases. In fact, this kind of gradient is obvious with human beings. The visible signs of aging (wrinkling, graying of hair, mottling and thinning of skin, loss of muscle tone, etc.) are highly visible, genetically controlled signals that do in fact make potential sexual partners less attractive to younger people. Aging adults are sometimes desperate to regain their youthful appearance. Attractiveness as a long-term partner is of course influenced by other things, and some of these things tend to increase as we age. For example wealth, power, status, and wisdom tend to accrue as we age. So the signs of aging are simply one factor in a complex equation of attractiveness. Even so, the predicted age-gradient of attraction is apparent in humans. Unfortunately for the theory, the gradient is missing in some other long-lived species. For example, female African bull elephants often prefer older bulls. Females in estrus tend to run away from younger male suitors. Clearly more needs to be done to clarify the applicability of the "benevolent reaper" theory to sexual attraction.
[131] Shackleford T and Larsen R. Facial attractiveness and physical health. *Evolution and human behavior*, 1999, 20, 71-76.
[132] Henderson and Anglin 2003. Page 354.
[133] Rhodes G, Simmons L, and Peters M. Attractiveness and sexual behavior: Does attractiveness enhance mating success? *Evolution and Human Behavior*, 2005, 26, 186-201. Page 187.
[134] Rhodes et al 2005. Page 196.
[135] Rhodes et al 2005. Page 198.
[136] Singh D and Young R. Body Weight, Waist-to-Hip Ratio, Breasts, and Hips: Role in Judgments of Female Attractiveness and Desirability for Relationships. *Ethology and Sociobiology*, 1995, 16, 483-507. Page 485.
[137] Singh and Young 1995. Page 502.
[138] Singh and Young 1995. Page 504.
[139] Townsend J and Wasserman T. Sexual attractiveness: Sex differences in

assessment and criteria. *Evolution and Human Behavior*, 1998, 19, 171-191.

[140] Townsend and Wasserman 1998. Page 175.

[141] Townsend and Wasserman 1998. Page 176.

[142] Townsend and Wasserman 1998. Page 189.

[143] Townsend and Wasserman 1998. Page 175.

[144] Townsend and Wasserman 1998. Page 174.

[145] The mathematical model these researchers put together suggests that characteristics reflecting health, intelligence, and physical symmetry all contribute to sexual attraction. Their results "imply that we are not attracted by good genes, but by a lack of bad genes. Sexual attraction is a force which counteracts genomic degradation." Morris R, Morris K, and Morris J. The mathematical basis of sexual attraction. *Medical Hypotheses*, 2002, 59, four, 475 – 481. Page 475.

[146] In 1975 a researcher named Zahavi introduced the idea that the very costliness of a feature like the peacock's tail ensures its honesty. He speculated that an animal may broadcast its high quality through a "handicap" that imposes a cost to the animal. Zahavi A. Mate selection - a selection for a handicap. *Journal of Theoretical Biology*, 1975, 53, 205-214.

[147] (Hamilton W and Zuk M. Heritable true fitness and bright birds: a role for parasites? *Science*, 1982, 218, 384-387).

[148] This kind of display should look familiar. Human males do the same kind of thing to advertise their status and wellbeing. They take women out to restaurants they can barely afford, drive fancy cars, wear nice clothes, show off. All these things are ways of saying, "I can easily handle these extravagances, baby. You could do worse than hook up with me."

[149] Saks L, Ots I, and Horak P. Carotenoid-based plumage coloration of male green finches reflects health and immunocompetence. *Oecologia*, 2003, 134, 301-307. Page 301.

[150] Saks et al 2003. Page 301.

[151] Rantala M and Kortet R. Courtship song and immune function in the field cricket Gryllus bimaculatus. *Biological Journal of the Linnean Society*. 2003, 79, 3, 503-510.

[152] Zala S, Potts W, and Penn D. Scent-marking displays provide honest signals of health and infection. *Behavioral Ecology*, 2004, 15, 2, 338-344.

[153] Biard C, Saulnier N, Gaillard M, and Moreau J. Carotenoid-based bill color is an integrative signal of multiple parasite infection in blackbird. *Naturwissenschaften*, 2010, 97, 987-995.

[154] Watve, MG and Sukumar, R. Asian elephants with longer tusks have lower parasite loads. *Current Science*, 1997, 72, 11, 885-889.

[155] DeBruine L, Jones B, Crawford J, Welling L, and Little A. The health of a nation predicts their mate preferences: cross-cultural variation in women's preferences for masculinized male faces. *Proceedings of the Royal Society B*, 2010, 277, 2405-2410. Page 2408.

[156] DeBruine et al 2010. Page 2409.

[157] Gangestad S and Scheyd G. The evolution of human physical attractiveness. *Annual review of anthropology*, 2005, 34, 523-548. Pages 533-534.

[158] The environment of a country with a low health index may be assumed to have greater incidence of parasites than a country with a higher health index.

[159] DeBruine L, Jones B, Crawford J, Welling L, and Little A. The health of a nation predicts their mate preferences: cross-cultural variation in women's preferences for masculinized male faces. *Proceedings of the Royal Society B*, 2010, 277, 2405-2410. Page 2408-2409.

[160] Lee A and Zietsch B. Experimental evidence that women's mate preferences are directly influenced by cues of pathogen prevalence and resource scarcity. *Biology Letters*, 2011, 7, 892-895.

[161] Gangestad S and Buss D. Pathogen prevalence and human mate preferences. *Ethology and Sociobiology*, 1993, 14, 89-96. Page 89.

[162] Park J, van Leeuwen F, and Stephen I. Homeliness is in the disgust sensitivity of the beholder: Relatively unattractive faces appear especially unattractive to individuals higher in pathogen disgust. *Evolution and Human Behavior*, 2012, 33, 569-577. Page 569-570.

[163] Park et al 2012. Pages 570, 575.

[164] Ridley M 2003. Page 72.

[165] Brookfield J. Population genetics. *Current Biology*, 1996, 6, 4, 354-357.

[166] For insights about population genetics in non-sexual species see Trevors J. Review: Bacterial population genetics. *World Journal of Microbiology and Biotechnology*, 1998, 14, 1-5.

[167] In 2000 researchers Vreeland et al reported recovering a living bacterium from a 250 million year old salt crystal. 250 million years is a long time in the past. That's before the present continents even existed! The researchers isolated the bacterium under exceptionally stringent sterile conditions to forestall any claim that it was simply a modern contaminant. The bacterium was found to be quite similar to a strain living today, containing genes that had been assumed to be "modern." The close similarity was used by critics to argue that it simply had to be a modern bacterium introduced during the recovery process. The experiment will likely have to be repeated multiple times to quell doubts. Original publication: Vreeland R, Rosenzweig W, and Powers D. Isolation of a 250 million-year-old halotolerant bacterium from a primary salt crystal. *Nature*, 2000, 407, 897-900. Critical response: Graur D and Pupko T. The Permian Bacterium that Isn't. *Molecular Biology and Evolution*, 2001, 18, 6, 1143-1146. See also: Maughan H, Birky C, Nicholson W, Rosenzweig W, and Vreeland R. The paradox of the "ancient" bacterium which contains "modern" protein-coding genes. *Molecular Biology and Evolution*, 2002, 19, 9, 1637-1639.

[168] Aufstad S. Making sense of biological theories of aging. In Bengtson V, Putney N, and Gans D (eds). *Handbook of Theories of Aging*, Second

Edition, 2008, Springer Publishing, 147-161. Page 156.

[169] Cao K, Blair C, Faddah D, Kieckhaefer J, Olive M, Erdos M, Nabel E, and Collins F. Progerin and telomere dysfunction collaborate to trigger cellular senescence in normal human fibroblasts. *The Journal of Clinical Investigation*, 2011, 121, 7, 2833-2844. Page 2833.

[170] Cao et al 2011. Page 2833.

[171] Norris J. Aging disease in children sheds light on normal aging. *University of California San Francisco News* (online), October 21, 2011.

[172] Mozes A. DNA 'telomere' length tied to aging, death risk. *US News Health*, Nov 8, 2012. Article online at http://health.usnews.com/health-news/news/articles/2012/11/08/dna-telomere-length-tied-to-aging-death-risk. Retrieved 3-16-13.

[173] Cao et al 2011. Page 2833.

[174] Winter C, Bouvier T, Weinbauer M and Thingstad T. Trade-offs between competition and defense specialists among unicellular planktonic organisms: the "killing the winner" hypothesis revisited. *Microbiology and Molecular Biology Reviews*, 2010, 42–57.

[175] Angly F, Felts B, Breitbart M, Salamon P, Edwards R, et al. The marine viromes of four oceanic regions. *PLOS Biology*, 2006, 4, 11, 2121-2131. Page 2121.

[176] Lopez-Garcia P and Moreira D. Viruses in biology. *Evolution: Education and Outreach*, 2012, 5, 389-398. Page 390.

[177] Weinbauer M, Wilhelm S. Gattuso J (Topic Editor). Marine viruses. Article in *Encyclopedia of Earth*. Eds. Cutler J. Cleveland (Washington, D.C.: Environmental Information Coalition, National Council for Science and the Environment). First published in the *Encyclopedia of Earth* December 21, 2006. Last revised Date November 5, 2011. Retrieved February 16, 2013 from http://www.eoearth.org/article/Marine_viruses.

[178] Winter C et al, 2010. Page 49.

[179] Angly et al, 2006. Page 2121.

[180] Bidle K and Vardi A. A chemical arms race at sea mediates algal host-virus interactions. *Current Opinion in Microbiology*, 2011, 14, 449-457. Page 450.

[181] Bidle and Vardi, 2011. Page 449.

[182] Weinbauer M, Agis M, Malits O, and Winter C. Bacteriophage in the environment. In McGrath S and van Sinderen D (eds). *Bacteriophage Genetics and Molecular Biology*, 2007, Caister Academic Press, Norfolk, UK. Pages 61-91. Page 81.

[183] Severin F, Meer M, Smirnova E, Knorre D, and Skulachev V. Programmed death of yeast Saccharomyces cerevisiae. *Biochimeca et Biophysica Acta*, 2008, 1783, 1350-1353. Page 1351.

[184] Huh J and Hay B. Apoptosis: Sculpture of a fly's head. *Nature*, 2002, 418, 926-928.

[185] See also Baehrecke E. How death shapes life during development.

Nature Reviews Molecular Cell Biology, 2002, 3, 779-787.

[186] Entezari M, Zakeri Z, and Lockshin R. *Apoptosis in developmental processes*. eLS John Wiley & Sons Ltd, Chichester.

[187] Koonin and Aravind 2002. Page 402.

[188] Sanmartin M, Jaroszewski L, Raikhel N, and Rojo E. Caspases: regulating death since the origin of life. *Plant Physiology*, March 2005, Vol. 137, pp. 841–847.

[189] Kirkwood B and Melov S. On the programmed/non-programmed nature of aging within the life history. *Current Biology*, 2001, 21, R701-R707. Page R704.

[190] Skulachev V. Aging as a particular case of phenoptosis, the programmed death of an organism (A response to Kirkwood and Melov "On the programmed/non-programmed nature of aging within the life history") *Aging*, 2011, 3, 11, 1120-1123. Page 1122.

[191] Kirkwood and Melov 2001. Page R705.

[192] Kirkwood and Melov 2001. Page R704.

[193] Kirkwood and Melov 2001. Page R705.

[194] Kerr J, Wyllie A, and Currie A. Apoptosis: a basic biological phenomenon with wide-ranging implications in tissue kinetics. *British Journal of Cancer*, 1972, 26, 239-257.

[195] Skulachev 2011. Page 1122.

[196] Severin et al 2008. Page 1352.

[197] Skulachev 2011. Page 1122.

[198] Sanmartin et al 2005. Page 841.

[199] The science fiction film Zardoz was released in 1974 and distributed by 20th Century Fox. It was written, produced, and directed by John Boorman. The male lead was played by Sean Connery, his second post-James Bond role.

[200] Jaskelioff M, Muller F, Paik J, Thomas E, Jiang S, et al. Telomerase reactivation reverses tissue degeneration in aged telomerase deficient mice. *Nature*, 2011, 469, 7328, 102–106.

[201] Life Extension Foundation. Life Extension® provides $2 million for new Age-Reversal Study. Online article retrieved 2-18-2013 from http://www.lef.org/featured-articles/Life-Extension-Provides-2-million-dollars-for-Age-Reversal-Study.htm?source=search&key=telomere.

[202] Jaskelioff M et al 2011. Page 102.

[203] Jaskelioff M et al 2011. Page 106.

[204] Jaskelioff M et al 2011. Page 102-103.

[205] Gomes N, Ryder O, Houck M, Charter S, Walker W, Forsyth N, Austad S, Venditti C, Pagel M, Shay J, and Wright W. Comparative biology of mammalian telomeres: hypotheses on ancestral states and the roles of telomeres in longevity determination. *Aging Cell*, 2011, 10, 761-768.

[206] Gomes et al 2011. Page 763.

[207] Gomes et al 2011. Page 761.

[208] Gomes et al 2011. Page 761.

[209] Life Extension Foundation 2013.

[210] The company is TA Sciences and their product is named TA-65®. The company website states "A double-blind, placebo controlled study of TA-65® showed improvements in: immune system, vision, male sexual performance, skin appearance, and more." The website also notes that use of the supplement has produced no reports of adverse effects.(Source: TA Sciences website http://www.tasciences.com)

[211] Hunt N, Hyun D, Allard J, Minor R, Mattson M, Ingram D, and de Cabo R. Bioenergetics of aging and calorie restriction. *Ageing Research Reviews*, 2006, 5, 125-143. Page 126.

[212] Dilova I, Easlon B, and Lin S. Calorie restriction in the nutrient sensing signaling pathways. *Cellular and Molecular Life Sciences*, 2007, 64, 752-767.

[213] Hunt et al 2006. Page 132.

[214] Riesen M and Morgan A. Calorie restriction reduces rDNA recombination independently of rDNA silencing. *Aging Cell*, 2009, 8, 624-632. Page 624.

[215] Vera E, Bernardes de Jesus B, Foronda M, Flores J, and Blasco M. Telomerase reverse transcriptase synergizes with calorie restriction to increase health span and extend mouse longevity. *PLOS One*, 2013, 8, 1, e53760, 1-13. Page 9.

[216] Vera E et al 2013. Page 1.

[217] Farzaneh-Far R, Lin J, Epel E, Harris W, Blackburn E, and Whooley M. Association of marine omega-3 fatty acid levels with telomeric aging in patients with coronary heart disease. *Journal of the American Medical Association*, 2010, 303, 250-7.

[218] Reported in Mackey D. New study shows omega-3s may lengthen telomeres. Agemarker.com, posted Oct 11, 2012. Retrieved 3-16-13. Original article: Kiecolt-Glaser J, Epel E, Belury M, Andridge R, Lin J, Glaser R, Malarkey W, Hwang B, Blackburn E. Omega-3 fatty acids, oxidative stress, and leukocyte telomere length: A randomized controlled trial. *Brain, Behavior and Immunity*, 2013, 28, 16-24.

[219] Mirabello L, Huang W, Wong J, Chatterjee N, Reding D, Crawford E, DeVivo I, Hayes R, and Savage S. The association between leukocyte telomere length and cigarette smoking, dietary and physical variables, and risk of prostate cancer. *Aging Cell*, 2009, 8, 405-413.

[220] See also Babizhayev M, and Yegorov Y. Smoking and health: association between telomere length and factors impacting on human disease, quality of life and life span in a large population-based cohort under the effect of smoking duration. *Fundamental and Clinical Pharmacology*, 2011, 25, 425-42.

[221] Tiainen A, Männistö S, Blomstedt P, Moltchanova E, Perälä M, Kaartinen N, Kajantie E, Kananen L, Hovatta I, and Eriksson J. Leukocyte

telomere length and its relation to food and nutrient intake in an elderly population. *European Journal of Clinical Nutrition*, 2012, 66, 1290-1294. Page 1290.

[222] Richards J, Valdes A, Gardner J, Paximadas D, Kimura M, Nessa A, Lu X, Surdulescu G, Swaminathan R, Spector T, and Aviv A. Higher serum vitamin D concentrations are associated with longer leukocyte telomere length in women. *American Journal of Clinical Nutrition*, 2007, 86, 1420-1425.

[223] Xu Q. Multivitamin use and telomere length in women, *American Journal of Clinical Nutrition*, 2009, 89, 6, 1857-1863. Page 1857.

www.ingramcontent.com/pod-product-compliance
Lightning Source LLC
Chambersburg PA
CBHW051317170526
45166CB00002B/585